T0238437

BestMasters

Weitere Informationen zu dieser Reihe finden Sie unter
http://www.springer.com/series/13198

Springer awards „BestMasters" to the best master's theses which have been completed at renowned universities in Germany, Austria, and Switzerland.

The studies received highest marks and were recommended for publication by supervisors. They address current issues from various fields of research in natural sciences, psychology, technology, and economics.

The series addresses practitioners as well as scientists and, in particular, offers guidance for early stage researchers.

Manuel Kroiss

Predicting the Lineage Choice of Hematopoietic Stem Cells

A Novel Approach Using Deep Neural Networks

 **Springer** Spektrum

Manuel Kroiss
Neuherberg, Deutschland

BestMasters
ISBN 978-3-658-12878-4   ISBN 978-3-658-12879-1   (eBook)
DOI 10.1007/978-3-658-12879-1

Library of Congress Control Number: 2016930594

Printed on acid-free paper

This Springer Spektrum imprint is published by Springer Nature
The registered company is Springer Fachmedien Wiesbaden GmbH

Abstract

We study the differentiation of hematopoietic stem cells using machine learning methods. This work is based on experiments focusing on the lineage choice of CMPs, the progenitors of HSCs, which either become MEP or GMP cells. Identifying the point in time when the decision of the linage choice is made is very difficult. As of now, the two biological markers GATA and FCgamma are used to identify these cells in an in-vitro experiment and then take them out for further expression analysis. However, prior results showed that an earlier detection might be possible and that the lineage choice seems to be several generations ahead of when the biological markers can identify the cells.

We present a novel approach to distinguish MEP from GMP using machine learning by using morphology features extracted from bright field images. Our model requires continuous measurements over one cell cycle with features describing the shape and texture of the segmented cell images. We test the performance of different models and focus on Recurrent Neural Networks with the latest advances from the field of deep learning. Two different improvements to recurrent networks were tested: Long Short Term Memory (LSTM) cells that are able to remember information over long periods of time, and dropout regularization to prevent overfitting. The best results were achieved by developing an extension of the LSTM model, using a bidirectional and deep-layered approach.

With our method, we considerably outperform standard machine learning methods without time information like Random Forests and Support Vector Machines. We were able to achieve a high accuracy of 80% after 60 hours of experiment time, while the biological marker was able to identify the cells only at 80 hours on average. In addition, we also trained our neural network with the expression level of the PU.1 transcription factor to study its potential use as a biological marker. By using PU.1, we were able to distinguish the cells with a high accuracy even earlier at 40 hours into the experiment. The classification accuracy can be improved to 90% by only taking high confident predictions. We measure the classification performance on different time-lapse movies with other experimental settings than the training data. Training the network on two movies takes about 3 hours on a GPU board, while making a prediction is in the range of 40 milliseconds. This allows us to potentially combine our method with an auto tracking approach that can be used in a live setup to identify the different cell types.

Zusammenfassung

In dieser Arbeit untersuchen wir die Differenzierung von hämatopoetischen Stammzellen mit Hilfe von Machine Learning Methoden. Die Ergebnisse basieren auf Experimenten, die sich mit der Zelllinienentscheidung von CMPs konzenrieren. CMPs sind die Progenitorzellen von HSCs, welche entweder zu MEP oder GMP Zellen differenzieren. Die Identifizierung des Zeitpunkts zu dem die Zelllinienentscheidung getroffen wird ist sehr schwierig. Bis zu diesem Zeitpunkt werden die zwei biologischen Marker GATA und FCgamma verwendet um die unterschiedlichen Zelltypen in einem in-vitro Experiment zu identifizieren und für weitere Expressionsanalysen zu extrahieren. Bisherige Ergebnisse im ICB haben jedoch gezeigt, dass eine frühere Erkennung der Zellen möglich sein könnte und dass die Zelllinienentscheidung schon einige Generationen vor dem Zeitpunkt liegt, an dem die biologischen Marker die Zellen identifizieren können.

Wir präsentieren einen neuen Ansatz mit neuen Machine-Learning Methoden um MEP von GMPs zu unterscheiden unter Verwendung von Form- und Texturmerkmalen die aus Hellfeldbildern von Mikroskopieaufnahmen berechnet werden können. Unser Modell benötigt kontinuierliche Aufnahmen der Zelle über einen Zellzyklus mit Merkmalen aus segmentierten Zellbildern. Wir testen die Leistung von verschiedenen Modellen und konzentrieren uns auf Recurrent Neural Networks unter Verwendung der letzten Fortschritte aus dem Deep-Learning Fachgebiet. Zwei verschiedene Verbesserungen für Recurrent Networks wurden getestet: Long Short Term Memory (LSTM) Zellen welche Informationen über lange Zeitspannen speichern können, und Dropout-Regularisierung um Überanpassung zu vermeiden. Die besten Ergebnisse wurden mit einer Erweiterung der LSTM Modelle mit einem mehrschichtigen und bidirektionalen Ansatz erzielt.

Mit unserem Modell können wir deutlich bessere Ergebnisse erzielen als gängige Machine Learning Methoden, die keine Zeitinformation verstehen wie z.B. Random Forests oder Support Vector Machines. Wir konnten eine hohe Genauigkeit von 80% nach einer Experimentzeit von 60 Stunden erzielen, während die biologischen Marker die Zellen erst nach etwa 80 Stunden erkennen konnten. Zusätzlich haben wir unser Neuronales Netzwerk auch mit den Expressionsmessungen des PU.1 Transkriptionsfaktors trainiert um dessen mögliche Verwendung als biologischen Marker zu untersuchen. Unter der Verwendung von PU.1 konnten die Zellen bereits nach 40 Stunden Experimentzeit mit einer hohen Richtigkeit erkannt werden. Die Genauigkeit beider Klassifikationsmethoden kann zusätzlich noch auf 90% verbessert werden, indem nur sehr sichere Vorhersagen genommen werden. In allen Fällen haben wir unsere Genauigkeit auf verschiedenen Mikroskopiefilmen gemessen, die nicht in den Trainingsdaten enthalten waren. Das Training des Netzwerks für zwei Filme benötigt etwa 3 Stunden auf einer Grafikkarte, während das Treffen einer Vorhersage nur etwa 40 Millisekunden benötigt. Diese kurze Zeit bei der Vorhersage würde es ermöglichen, die Methode mit einem Ansatz zur automatischen Zellverfolgung zu kombinieren, welcher während einem laufenden Experiment verwendet werden könnte.

Acknowledgements

I would like to thank Florian Büttner and Felix Buggenthin for their excellent supervision and guidance during my thesis. Especially taking the time for the long discussions on how to address the different problems, as well as the valuable comments on the thesis and presentations. I would also like to thank Fabian Theis for his valuable input and feedback.

Contents

Figures

Tables

1 Introduction

1.1 Machine Learning

The focus of *machine learning* is to develop algorithms that can learn from data [1]. The traditional approach to solving a data-driven problem is to build an algorithm that describes a systematic procedure of calculations specifically for the problem [2]. However, data from experiments or observations is often noisy and building such an algorithm for high-dimensional data is not feasible in most scenarios. Machine learning offers a solution to this problem. Instead of looking at the observations by hand and manually finding patterns in the data, machine learning "gives the computers the ability to learn without being explicitly programmed" [3].

The power of a machine learning model can be measured by its ability to learn and generalize [4,5]. Generalization describes the ability of the model to perform accurately on unseen observations. Achieving a good balance between generalization and capability to learn is key in machine learning models. For parametric models, increasing the learning capability of a model usually leads to a better performance on the training data. However, often times the increase in performance stems from the model learning the individual examples by memorizing individual training examples. In this case, the generalization degrades and the performance for unseen examples drops [6].

There are two different approaches to learning from data. The general case is unsupervised learning, where the model is trained on a set of unlabeled examples [7–9]. This means, that the training data is unordered and there is no prior knowledge about the underlying structure of the data. The model aims to find common patterns in the data that best helps distinguish the individual examples from each other.

A different problem statement is supervised learning [5,10]. Instead of unordered data, the model also gets a label for each example in the set of observations. In the case of classification [11,12], the focus is not to distinguish individual examples, but different groups of labels. The other supervised learning problem is regression. In statistics, a supervised classification problem is often solved with a logistic regression model, since the probability output can be interpreted as the probability for class-membership [13,14].

Several mathematical methods like Support Vector Machines [15] or Random Forests [16,17] have been developed for supervised classification. Unfortunately, the level of abstraction these models can handle is usually very limited. In most cases, it is required to build an algorithm that extracts meaningful features from the raw data before the data can be learned [18,19].

A machine-learning problem that has received a great deal of attention in the recent years is image classification. Building features that describe an image requires manual analysis and lots of effort. In most cases, images are very noisy and different from each other, and there are no universal image descriptors that fit every case [20,21]. The process of building features is very problem dependent and often this work cannot be transferred to different areas. This becomes especially difficult when the images cannot be classified manually and it is unclear if the label can be deduced from the data.

Instead, a different approach is taken, starting with a mediocre representation of the data with features that describe it more in a general or in an imprecise manner. The idea is to then use these low-level features and build a model that looks at many examples to find similarities and patterns in these observations. Just like the human brain, the algorithm learns abstraction [22]. The difficulty remains in constructing such a model, but once it is built, it can be applied to all kinds of problems with numerous training examples. With the beginning of the deep learning era, models that can perform a high level of abstraction were introduced.

1.2 Deep Learning

During recent years, *deep learning* has been a topic of great interest in the field of artificial intelligence. Deep learning can be described as a way of analyzing information with a layered structure where the goal is to gain higher levels of abstraction by introducing more layers in the model [23–25]. These layered models assume that the information can be organized in different levels of detail with a hierarchical structure [26]. The success of this approach has made it an essential part of machine learning and also the gold standard in solutions for problems like handwriting analysis [27], image classification [28] and speech recognition [29].

The essence of deep learning is that the need for designing special features that perfectly describe an observation becomes less important. Instead, one tries to design models that can build the necessary features themselves from a mediocre representation or even raw input of images or audio signals. Given a low-level representation of the data, the algorithm learns to build layers of features that become increasingly meaningful until the solution to the problem becomes obvious. Essentially, building features can be interpreted as a way of making the data more abstract.

A model that lends itself perfectly to the idea of a layered representation learning is the Artificial Neural Network [30,31]. Therefore, deep learning commonly refers to learning neural networks with many layers [32]. The first deep-layered feed-forward networks were published in mid-2000 by G. Hinton in Toronto and simultaneously in Montreal using a different approach by J. Bengio [23,25]. Both models rely on first training an unsupervised model that learns an increasingly abstract representation of the data by stacking hidden layers of neural networks. Each layer was trained separately to find a good representation of the layer below given the amount of units it has available. After pre-training the layers, one would attach the labels for supervised classification on top of the last layer and then fine-tune the free parameters of the model slightly to fit the labels rather than a general representation of the data.

The introduction of the unsupervised pre-training technique allowed further investigation into the characteristics of stable deep networks. It was found that the stability and generalization of a network highly depends on the range of the parameters [33]. Later studies have shown that neural networks can be trained without any pre-training step when initializing the parameters of networks properly [34]. This allows the immediate training of a supervised deep-layered model.

Although deep nets have been topping most machine learning competitions, there has been surprisingly little application in bioinformatics, a field with many problems that have large undiscovered datasets [35–37]. The newest methods in machine learning and especially representation learning could be very beneficial to many open problems in biology. A topic of interest in medicine with great potential is research in stem cells.

1.3 High-throughput time-lapse microscopy of murine stem cells

The benefits of stem cell research are great with possible therapies for diseases such as cancer, cardiovascular disease or dementia [38,39]. Stem cells are defined as biological cells that possess the capacity of *self-renewal* and have the *potency* to differentiate [40,41]. This means that stem cells are able to produce daughter cells with the same properties as their own, but also give rise to more specialized cell types.

In our work, we focus on hematopoietic stem cells (HSCs), which are blood stem cells that can be found in the bone marrow of adults [42]. They are responsible for rebuilding cells of the blood and immune system. Studies of HSCs resulted in clinical treatments for leukemia [43] and immune system disorders [44].

The different classes of hematopoietic stem cells can be categorized by their potency and structured in a tree [42,45] (see Figure 1-1). At the root of the hierarchy are the HSCs, which have the ability to self-renew and can give rise to hematopoietic progenitor cells (HPCs) that are more specific and can only divide a limited number of times. Our studies focus on the progeny of the common myeloid and lymphoid progenitor (CMP) cells, which are either megakaryocyte-erythroid progenitors (MEP) or granulocyte-macrophage progenitors (GMP) cells. MEP cells are the predecessor of red blood cells and megakaryocytes, while GMP cells become granulocytes or macrophages.

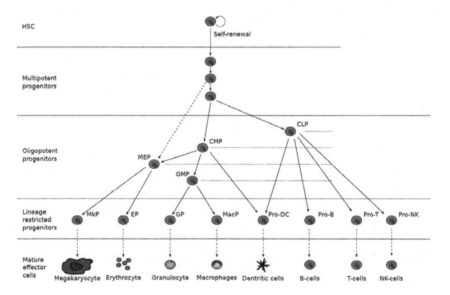

Figure 1-1: Hierarchy of hematopoietic stem cells, adapted from [45]

As of now, there is little knowledge about the molecular processes, which are involved in the differentiation of hematopoietic progenitor cells [46]. Methods to measure the molecular processes, such as expression profiles of gene regulation, require the cells to be lysed and destroyed in the analysis. As the time point of differentiation is not known, a snapshot analysis at the right point in time is very difficult. If the analysis is performed slightly too soon or too late, information about what caused the differentiation might be gone.

Therefore, non-invasive methods for continuous long-term study of the cells are required [47]. A very promising approach ideally suited for this purpose is live-cell imaging of in-vitro experiments [48–50]. As taking microscopic images of cells can kill or change the behavior of the cells, the images have to be taken in short intervals of several minutes. Repeated images are taken over several days, resulting in a long-term microscopy movie of the cells.

Markers used in live-cell imaging are either fluorescent proteins that are encoded in the genetic sequence of the cell, or antibodies that bind to specific molecules of the cell [47,51,52]. The fluorescence levels of the cells are measured using cell imaging of the specific fluorescent wavelengths. In the case of MEP and GMP, biological markers are known that correlate with the differentiation. These markers allow the labeling of the cells.

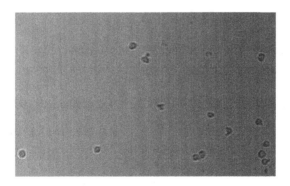

Figure 1-2: Example of a bright field image from long-term microscopy [53]
The cells are the black objects with the white halo in the images. The other small particles are dirt.

When looking at the tracked genealogies, we observe that most siblings will differentiate to the same type, either all MEP or all GMP. As shown in Figure 1-3, the left sub tree has only MEPs and the right ones only GMPs. This suggests that the decision whether the cell becomes GMP or MEP seems to be several generations ahead of when the biological markers become active. This raises the question whether there is a way to distinguish the cells before the onset of the currently used markers.

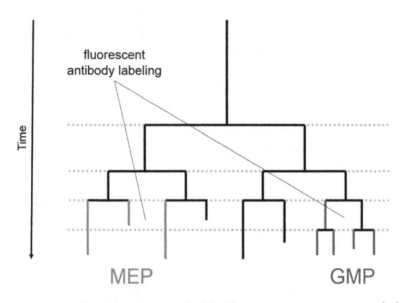

Figure 1-3: Labeled tracking tree of HSC differentiation to MEP or GMP [54]
Each line is branch in the tree is one linage, consisting of several cell cycles. When the cell divides, the branch is split. Green are cell cycles with an active MEP marker, while blue are cell cycles with an active marker for GMPs.

Two potential sources of information that have not been used for labeling so far are the bright field images and the expression levels of other biological markers. Usually, the bright field images are only taken for tracking purposes. However, in preliminary studies [55] it was shown that morphological features of the cells could potentially be used to distinguish MEPs from GMPs. Our goal is to further investigate these ideas using innovative approaches of deep learning.

1.4 Problem Statement

The goal of this thesis is to use deep learning methods to distinguish between MEP and GMP cells. As training data, we use sequences of bright field images for each cell over one cycle. The cells are manually labeled, if the fluorescent level of the marker passes a certain threshold. This problem is called supervised binary time series classification.

We want to address this problem using deep recurrent neural networks. Despite the advances of these models in the last years, taking raw images as input is still too complex when dealing with time series data. Therefore, we use features describing the shape and texture of segmented cell images. Nevertheless, these models still have to achieve a high level of abstraction as they are presented with temporal input and have to find patterns and similarities in time.

We want to apply our classifier to other time-lapse movies and predict the label unknown cells. As the tracking tree suggests that the lineage choice of the cells is earlier than the biological markers become active, we hypothesize our approach might be able to distinguish the cells at earlier stages.

1.5 Structure of the thesis

In the first three of chapters, we give a general introduction into the techniques we are using. First, we present basic Artificial Neural Networks and Feedfoward models that can handle one-dimensional input. Afterwards we cover the structure as well as problems of Recurrent Neural Networks and possible solutions on how to address these issues. The most important part of a neural network is the training process where we focus on the two classic optimization techniques that are still the most commonly used approaches and different regularization techniques that prevent overfitting of the networks when learning. Finally, we show and discuss the results of our studies and conclude with ideas on how to continue this work.

2 Introduction to deep neural networks

2.1 Feedfoward Neural Networks

Artificial Neural Networks (Anns) are mathematical models for machine learning, that were inspired by the structure of the central nervous system in animals or humans [56,57]. The human brain is still the most powerful information processing unit for complex tasks like perception, recognition and abstraction. By mimicking the biological structure of neurons and synapses, the hope is to gain a similar capability for understanding information and apply it to solve simple tasks.

The basic structure of an ANN is a network that consists of nodes and weighted directed connections between these nodes. Biologically speaking, the nodes, also called units, represent neurons and the weighted connections represent the strength of the synapses between the neurons. To process information, a subset of the nodes is activated, the so-called input neurons, which in turn activate other neurons by transferring their signal. The signals are fed through the network until an output unit is reached, which is the result of our information processing and hopefully the desired solution.

The goal of the learning process of an ANN is to use a training set of labeled examples and minimize a task-dependent *objective function* in order to achieve a good performance on the test set, which are the examples that the model has not seen before. In most cases, the performance can be measured by an error function that compares the predicted with the actual labels of the data. For parametric models such as neural networks, the error can usually be minimized by iteratively adjusting the free parameters. The ability of an algorithm to transfer performance from the training set to the test set is called *generalization* and is a key part for learning models.

Although it is now known that the models are far from the complexity of a real brain, the methods have enjoyed continued popularity for pattern recognition and regression tasks. Especially in recent years, with the advent of deep learning [58], a field in neural networks where new ideas are employed to achieve a very deep structure with many units, these models have become very powerful.

2.1.1 Perceptron

The perceptron [57] is the basic building block of an ANN and generally seen as the first generation of neural networks. It is an algorithm for supervised binary classification, which is able to learn the relation of data x and target z of a given set of trainings examples. The classifier works by predicting a binary output from the data and comparing this prediction with the correct class label. The free parameters of the model are then iteratively adjusted in order to close the gap between predicted target and actual target in the training set.

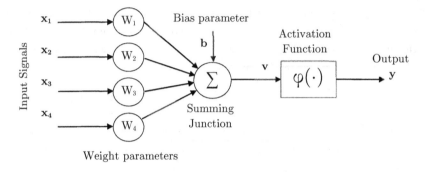

Figure 2-1: Structure of a Perceptron [54]

The features of an observation x act as input signals to the perceptron (see Figure 2-1). Each signal x_i is weighted by a free parameter w_i, which represents the influence of the input signal on the output. The weighted signals are summed up and a bias b is added, which can act like an offset to the linear combination. Afterwards, the combined value is transformed by an non-linear activation function φ. To produce a binary output value between 0 and 1, the *sigmoid function* σ (see Figure 2-2) (also called *logistic function*) is used as an activation function.

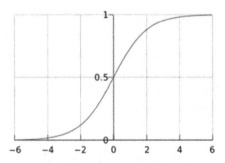

Figure 2-2: Sigmoid activation function [59]

The complete computation of the output y can be described with the following equation:

$$y = \sigma \left(\sum_{i=1}^{N} w_i x_i + b \right)$$

The goal of the classifier is to minimize the errors in prediction. Therefore, the resulting output y is compared with the supposed output z for every data-target pair (x, z). To quantify the amount of errors, the *zero-one loss* function can be used:

$$L(\theta, D) = \sum_{i=1}^{|D|} I_{y_i \neq z_i}$$

$$I_x = \begin{cases} 1 & \text{if } x \text{ is True} \\ 0 & \text{otherwise} \end{cases}$$

where θ are the free parameters and D is the set of training examples.

However, the zero-one loss function is not differentiable and it would be very expensive for large models and many data pairs. Thus, the *negative log likelihood* is used as a loss function:

$$NLL(\theta, D) = - \sum_{i=1}^{|D|} \log P(Y = z_i | x_i, \theta)$$

This loss function is iteratively minimized, by calculating the gradient of the loss and updating the free parameters accordingly. Optimization techniques are described in the following chapters.

The perceptron can tackle binary classification problems. However, modeling a multi-class classification problem is a bit more difficult. Multiple perceptrons have to be used, where each of them models a yes or no answer for one of the classes. This is equal to having a perceptron with multiple output units, as the output nodes are independent. To obtain a probability value for each class, the *softmax function* is applied to the final output units of the network:

$$softmax(x)_i = \frac{e^{x_i}}{\sum_{k=1}^{K} e^{x_k}}$$

Interestingly, the loss function stays the same even for the multiclass case.

2.1.2 Limitations of a Perceptron

If the right features are chosen by hand, a perceptron can learn almost anything. However, when using the wrong features, it can be very limited.

Consider an example with two binary features and the goal is to perform binary classification. Positive examples (same) are (1,1) and (0,0). Negative examples (different) are (0,1) and(1,0). We want to do a check for equality. If both features are equal, either both one or both zero, then the observation is classified as positive. Otherwise, if the binary features are different, the classification should be negative.

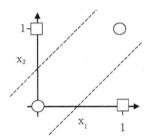

Figure 2-3: Limitation of a perceptron [54]
the four data points of two different classes cannot be separated linearly

This example is impossible to solve with a perceptron as shown geometrically in Figure 2-3. The weight plane of the perceptron can be seen as a line and it is obvious that is impossible for this line to separate the four data points properly. A solution to this it is to stack perceptrons and thereby separate the data linearly with two lines.

However, the case becomes more difficult for a problem where the points have to be separated in a non-linear way. In this scenario, stacking perceptrons with linear units is not enough as they can still only perform a linear separation of the data set. A combination of non-linear functions and multiple layers of perceptrons is needed, which we will discuss in the next chapter.

2.1.3 Multilayer Perceptron

The *Multilayer Perceptron* (MLP) [22,30] architecture, is as the name suggests, the result of stacking perceptrons in a layer wise manner. A single perceptron is very limited in its ability to model data and cannot distinguish input that is not linearly separable. This problem is tackled by stacking perceptrons as depicted in Figure 2-4. Given enough perceptron units, the MLP is able to approximate any boundary function and thus has a universal approximation capability [60,61].

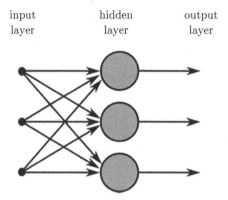

Figure 2-4: Structure of a Multilayer Perceptron [62]

Each layer of the MLP is connected by using the outputs of the previous layer as inputs for the next one. Instead of the sigmoid function, it is common to use the *tanh function* as the activation function for the hidden layers. While the hyperbolic tangent is a sigmoidal function and just a linear variation, it was shown that neural networks using tanh layers learn better [33]. The problem is that the sigmoid function is not centered on zero, which introduces a bias that causes the activation to saturate towards one rather than towards zero.

Depending on the problem statement, it is also possible to change the activation function of the last layer. For binary classification, the sigmoid function is used. For logistic regression, where the target is a probability value, the problem is the same as binary classification and sigmoid function can be used again. However, for regression of a real value, e.g. predicting the size of a loan in banking, the identity function can be used and no transformation is required in the last layer.

Using matrix notation, the equations of an MLP can be denoted as follows:

$$h = \tanh(W_{xh}\, x + b_h)$$
$$y = \varphi(W_{hy}\, h + b_{by})$$

where h is a vector with one element for each unit of the hidden layer, and y is a vector with one element for each output unit. φ is the activation function of the output layer, either the sigmoid function, the softmax function or the identity function depending on the type of problem. W_{xh} is a weight matrix of size $\#input * \#hidden$ and W_{hy} has the dimensionality $\#hidd \quad * \#output$. The bias vectors b_h and b_y have the same size as h and y respectively. A weight matrix is multiplied with a column vector by the matrix-vector product. The equation for the hidden layer can be repeated for additional hidden layers, using the output of the previous one as input for the next one. Note that for each layer, a separate set of free parameters W and b is needed.

2.1.4 Backpropagation

To calculate the gradient of the loss function in a neural network, a technique called *backpropagation* [22] is used. The partial derivative of the loss function E results from using the chain rule:

$$\frac{\delta E}{\delta w_{ij}} = \frac{\delta E}{\delta o_j} \frac{\delta o_j}{\delta v_j} \frac{\delta v_j}{\delta w_{ij}}$$

The change of a weight w_{ij} connecting neuron i and j can be calculated as follows:

$$\Delta w_{ij} = -\frac{\delta E}{\delta w_{ij}} = \eta \delta_j x_i$$

$$\delta_j = \begin{cases} \varphi'(v_j)(z_j - o_j) & \text{if } j \text{ is an output unit} \\ \varphi'(v_j) \sum_k \delta_k w_{jk} & \text{if } j \text{ is a hidden unit} \end{cases}$$

Where v_j is the linear combination of the input to a neuron j with bias before applying the activation function and o_j after applying the activation function φ. z_j is the supposed output of the network.

2.2 Recurrent Neural Networks

A Recurrent Neural Network (RNN) is a different type of ANN architecture. So far, we have only discussed feed-forward architectures like the MLP that do not contain any cycles. With the introduction of RNN, directed cycles are allowed between the units in the network. This creates an internal state, which enables the network to remember information and therefore process temporal information.

Instead of a single vector of features per observation, each observation x can now be a series of feature vectors, representing a time series. Note that each of these feature vectors must have the same length, but the different observations of a data set can have different time lengths. Each time step of an observation is fed iteratively into the network.

2.2.1 Basic RNN

The basic structure of an RNN (see Figure 2-5) usually contains one recurrent hidden layer. Each unit of this recurrent layer is fully connected with every unit in the output layer and also with every unit in the recurrent layer itself. When processing a time series, feature vectors are fed into the network in a stepwise manner. For each time step, the new value of the output units and a value for every hidden unit is calculated. In the next time step, this value for the hidden unit is combined with the new incoming input, resulting in a new output and hidden value. This is how the network is able to remember and understand temporal information.

input   recurrent   output
layer   hidden layer   layer

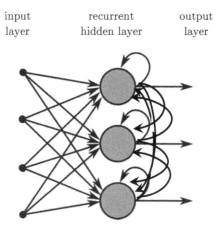

Figure 2-5: Structure of a Recurrent Neural Network [62]

This structure can be described with the following equations:

$$h_t = \tanh(W_{xh}\, x + W_{hh}\, h_{t-1} + b_h)$$
$$y_t = \varphi\big(W_{hy}\, h_t + b_y\big)$$

Note that the term $h_{t-1}\, W_{hh}$ is the main difference between the equations of a MLP (see chapter 2.1.3). W_{hh} is the weight matrix for the hidden units of size *#hidden* $*$ *#hidden* and h_{t-1} are the values of the hidden units from the previous time step.

The result of processing a time series is a vector with one output y_t for each time step. In the case of a sequence classification, where we are only interested in the label of the whole time series, it is common to simply take the mean of the output units for each time step and obtain an average prediction. The same kind of loss functions as for MLP are then applicable to the RNN.

In theory, given enough neurons and time, RNNs can model and calculate anything that can be computed by a computer. The problem however is that these RNNs are very difficult to train.

2.2.2 Vanishing Gradient Problem

A major problem when training a RNN is the vanishing gradient [33,63,64]. The sensitivity of the RNN to a time point decays exponentially as time passes and as more steps are processed. New inputs overwrite the information in the hidden units and the network forgets the first inputs.

The problem can be illustrated when unfolding the RNN over time (see Figure 2-6). In the figure, the shading indicates the sensitivity of the network to the input signal. Over time, the gradient vanishes and after ten or more steps, the information is gone completely.

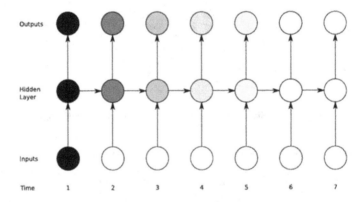

Figure 2-6: Problem of the vanishing gradient for RNNs, adapted from [65]. The shading indicates the receptiveness of the network to the input information. The receptiveness decays exponentially as new input overwrites the information stored in the hidden units and the network forgets earlier information.

In the forward pass of a neural network, the common activation functions like the logistic function limit the activities between fixed intervals, which prevents the activity from exploding. For the backward pass however, the system is completely linear. This means that when doubling the error derivatives in the final layer, all derivatives of the layers before will double as the error is back propagated.

Because a neural net is a linear system during backpropagation, it has the same problem as linear systems when iterated: their numbers grow or die exponentially. On the one hand, very small gradients will die and the information is lost since it is too small. On the other hand, big gradients will explode and the information will be wiped out. Generally, this behavior causes the information of the gradients several time steps before to be lost.

This problem does usually not occur in feed-forward networks, since feed-forward nets have only a couple of layers. For RNNs however, e.g. if the training sequence has 100 time steps, the RNN is essentially a network with 100 layers. Thus, the resulting gradient is to the power of 100. There are a few ways to handle this problem, either on the architecture side or on the trainings side, which will be discussed in the following chapters.

2.2.3 Long Short-Term Memory

A very good way to address the problem of the vanishing gradient is the use of Long Short-Term Memory (LSTM) cells [65–67]. Instead of a normal neuron, which just passes the values through to the next neurons in the network, LSTM cells have a special memory block inside where they can store information.

The dynamic state of the network can be seen as a short-term memory. The idea of the LSTM cells is to use this model of a short-term memory and create a long-term memory on top of it. This means, building a module that can store information. A crucial point of this model is that it is separated from the rest of the network and new input does not interfere with the memory.

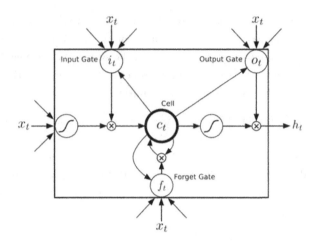

Figure 2-7: Structure of a Long Short Term Memory cell, adapted from [67].
The cell has an additional memory store where it can read, write or forget
information. These operations are initiated by gate functions that depend on new
input values coming into the cell.

The memory block is separated by logistic gates from the rest of the network (see
Figure 2-7). There are three kind of gates: input, output and forget gates. Whether to
open one of these gates for a cell is determined by the rest of the network. When the
network wants to store information, it opens the input gate and new information floods
into the memory block.

The information stays there as long as the forget gate is closed. Again, the rest of the
network determines the state of the logistic forget gate. To read the information, the
rest of the network simply opens the output gate. This causes the stored value to flood
out of the cell and affect the rest of the network.

The memory block can be thought of as a value storage unit. It can store the exact
value of a number. This number remains unchanged as long as the input and forget
gates are closed. If the network wants to forget the value, it can simply open the forget
gate and the number is overwritten by zero. New information can then be stored again
by opening the input gate.

The structure of the LSTM can be formalized as follows:

$$i_t = \sigma(W_{xi}x_t + W_{hi}h_{t-1} + W_{ci}c_{t-1} + b_i)$$

$$f_t = \sigma\big(W_{xf}x_t + W_{hf}h_{t-1} + W_{cf}c_{t-1} + b_f\big)$$
$$c_t = f_t c_{t-1} + i_t \tanh(W_{xc}x_t + W_{hc}h_{t-1} + b_c)$$
$$o_t = \sigma(W_{xo}x_t + W_{ho}h_{t-1} + W_{co}c_t + b_o)$$
$$h_t = o_t \tanh(c_t)$$

where i_t is the output gate, f_t the forget gate, c_t the value of the memory block that is stored, o_t the output gate and finally h_t the hidden units.

It was shown that the LSTM cells can remember information for several hundred time points. This means they do not suffer from the vanishing gradient problem. The task they have shown to be best at is understanding handwriting [68]. Training the LSTM network is the same as for RNNs and can be performed with gradient descent methods (see chapter 2.3). The structure of an LSTM network is usually similar to a RNN and only has one recurrent layer.

2.2.4 Bidirectional RNN

An extension of the RNN is the introduction of a bidirectional structure (see Figure 2-8) [69]. Instead of only using one RNN to process the time series in the forward direction, a second RNN is used that processes the data in the backward direction. At each time step, the information from the forward and backward direction is combined to make the prediction for the network output.

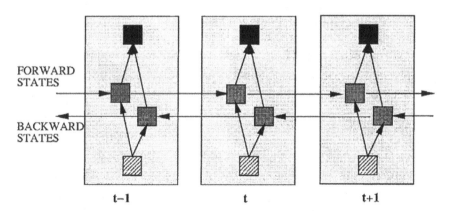

Figure 2-8: Bidirectional RNN unfolded in time, adapted from [69].
At each time step, the network has the full sequence information available from reading the sequence in the forward and backward direction. The output layer merges the information from the two recurrent networks.

The main advantage of this structure is that the model has seen the complete time series at every time step, the first part in the forward and the second part in the backward direction. Therefore, the output at each time step is much more meaningful. However, using the complete sequence for classification makes live prediction impossible.

When only using a forward RNN, the model makes a prediction about a time step without having seen the later time steps. This issue could be addressed by introducing a small window into the future where the RNN has access to the next few time steps. This however would introduce many additional weights. A better solution might be to introduce a time delay for the output layer. However, the length of the window would still be a very sensitive parameter that the user has to guess for every problem manually.

Especially in the case of sequence classification, the first outputs of a basic forward RNN are questionable and very problem dependent, since the RNN has only seen very few time points so far. Weighting the outputs of the model with increasing importance towards the end of the sequence is possible, but the important information could be in the first time steps and since the net slowly forgets the information over time, this would be very counter intuitive. Instead, using a bidirectional RNN is a much better solution and less arbitrary.

2.3 Training Neural Networks

2.3.1 Input normalization

Neural networks are generally very robust to unnecessary features. However, when the dimensionality of the input becomes too big, learning the weights becomes increasingly difficult. Further, if the data set is small, the network becomes prone to overfitting on the unnecessary features.

An important preprocessing step that should be applied to every data set is to normalize the mean and the standard deviation of the input. Each feature should have a mean at zero and a standard deviation of one over the whole data set. This preprocessing step does not change the information in the data set, but ensures that the input is in a better range for the initial parameter values for the network [70]. Learning becomes easier and the trainings process more stable.

2.3.2 Parameter Initialization for RNN

In recent work [71] it was shown that the initialization of a recurrent neural network plays a very important role in the success of the trainings process. If the parameters are not initialized properly, the model is likely to end up in local minima, where it has not achieved any generalization, but only learned to fit noise of the training set. One might argue that a good optimizer should be able to escape local minima, but if it starts far off from a good solution, then it is very unlikely that the optimizer will ever reach a good solution because there are many steps in between where the error value does not improve.

An important part of the RNN initialization is the spectral radius of the hidden-to-hidden weight matrix. The spectral radius of a matrix is the eigenvalue that has the largest absolute value. It was shown [72] that this has a great influence on the dynamics of the hidden units when using a tanh activation function. If the spectral radius is smaller than one, then the RNN will tend to quickly forget information. When it is much bigger than one, then the information is saved for longer periods, but also experiences chaotic behavior, which can lead to exploding gradients. As discussed in chapter 2.2.2, exploding gradients often wipe out existing information, so the network tries to remember too much information. However, if the spectral radius is only slightly bigger than one, the information is remembered without experiencing exploding gradients. This means the network remembers just as much information as it can handle. Therefore, it was suggested to use 1.1 as an optimal value for the spectral radius.

Further, it was also found that the initial scale of the weights are very important as well. For the input-to-hidden units, it depends on whether there are many uninformative features in the data set. Generally, if the initial scale of the weights is set too high, then the information is overwritten easily and features that are not important will wipe out other useful information. However, when the initial scale is too low, the learning process tends to be slow, which might not ever lead to a proper solution. As a guideline, using a Gaussian distribution with a standard deviation of 0.001 for the input-to-hidden weights is recommended. However, if it is certain that all features are useful, the value for the standard deviation can be increased up to 0.1 to speed up the learning process.

Parameter type	Scale
input-to-hidden	N(0, 0.001)
hidden-to-hidden	Spectral radius of 1.1
hidden-to-output	N(0, 0.1)
hidden-bias	0
output-bias	average of outputs

Table 2-1: Optimal initial parameter values for RNNs

In addition to the scale of the parameters, the centering of the input and output is also crucial. As suggested in chapter 2.3.1, the mean of the input should be zero with a standard deviation of one. Additionally the output should also have a mean of one and this can be fixed by setting the initial bias value for the output units as the average of the correct output values.

2.3.3 Parameter optimization methods

Unfortunately, there is no best optimization method for training neural networks. The issue with training them is that there are many different scenarios where different aspects of the learning process are important [73].

First, the structure of neural networks is often very different. There are deep networks with many layers, which makes them hard to train. The optimizer has to be very sensitive to small gradients to detect necessary changes in the first layers. The same is the case for recurrent networks, which have even more layers if the structure is unfolded the structure. Time sequences can have more than thousand steps, this means when reaching the end of the sequence the information has gone through more than thousand layers. To find the correct solution for the problem, the RNN often has to find relations between information that happened over these long periods. A very different learning scenario occurs when having shallow networks with only few layers but many parameters. Optimizing the parameters the same way as for deep networks would drastically overfit the data. Instead, the parameters are not trained very sensitively and the training is stopped early as long as the networks has a good generalization.

The other issue is that the problems vary widely in their type of input. For pixels values, the input is very dense. Other times for e.g. binary input for words, the input is usually very sparse. This difference in input has a huge impact on the expected values for the weights. For highly informational features like words, the weights should be very sensitive. However, if the input is dense like for pixels or speech signals, the network has to find a good abstraction of the data first, which generally requires small evenly distributed weights.

To decide what optimization technique to use in which scenario, we also have to look at the amount of examples in our training set. As a general guideline, the different algorithms can be categorized as follows [73]:

Small dataset ($<$ 10000 examples)	Big redundant dataset
Conjugate gradient	Minibatch-SGD with momentum
LBFGS	Rmsprop
Adaptive learning rates	
RPROP	

Table 2-2: Different optimization methods depending on the size of the data set

While the algorithms for small datasets are usually full batch learning methods, the ones for big redundant datasets are using batch learning. A special case is the Hessian-Free optimization [74] technique which was tailored specifically for very deep or recurrent neural networks. This algorithm is usually only applicable to small data sets due to speed constraints, but makes use of mini batches for learning.

2.3.3.1 Stochastic Gradient Descent

Gradient descent is a technique for updating the free parameters of a model. The optimizer iteratively takes small steps on the error surface defined by a loss function with respect to the parameters by using the gradient.

For *ordinary gradient descent,* a stepwise iteration is performed. In each step, the error of the loss function and the gradient of the weights is calculated, with respect to the loss on the whole trainings data set. Using the gradient calculation from backpropagation (see chapter 2.1.4), the weights are updated with a factor according to the learning rate.

Stochastic gradient descent (SGD) has the same approach as ordinary gradient descent, except that the gradient is not calculated for the full data set in each step. Instead, the gradient is estimated from one example to proceed more quickly. Not only does this procedure save computation time, but it also adds randomness to the descent that lets the optimizer escape local minima more easily.

The main hyper-parameter for gradient descent based methods is the learning rate. This factor determines the step size that is taken in the direction of the gradient. Unfortunately, this parameter is highly problem dependent and needs to be adjusted during training. The later we are in the trainings process, the smaller the adjustments to the parameters should be. This can be solved by introducing an additional hyper-parameter for momentum, which controls the decay of the learning rate after each iteration through the training set (also referred to as epoch).

A common variation of SGD is the *mini-batch SGD* that estimates the gradient from a few trainings examples instead of only one. This results in a more stable learning process. It is interesting to note that increasing the size of the batch from one to five results in a drastic reduction of the variance and a more stable learning process. However, increasing the batch size beyond 20 usually does not help and the improvement fades rapidly.

2.3.3.2 Resilient Propagation

Resilient Propagation (RPROP) [75] is a full batch learning method. It is very robust and a good general-purpose optimization method for both feed-forward and recurrent neural networks.

Gradients in neural networks often vary widely in their magnitude. Some of the gradients are huge and others are very small. This is especially true in the case of the RNNs as discussed in chapter 2.2.2. The central idea of the algorithm is to only use the sign of the gradient and completely ignore the magnitude. A problem that occurs if we only take the sign is that it is very hard to choose a suitable learning rate and momentum parameter that fits all parameters well, as they vary greatly in size. The solution of the RPROP algorithm is to introduce an individual learning rate, in this case called step size, for every parameter.

In the algorithm, the step size adapts over time. When the parameter was moved in the correct direction, the step is increased multiplicatively in size. Otherwise, the step size is decrease multiplicatively. The direction was right, if the sign of the last two gradients agree.

Good example values for the factor of the positive increase in step size is 1.2 and for the negative decrease 0.5. To ensure that the step sizes are limited, a common value for the lower bound is 10^{-6} and 50 for the upper bound. For the initial step size, 0.0125 is proposed as a suitable value. The hyper-parameters can be chosen problem dependent, but they are very robust and usually have little impact. This is different from momentum SGD where the learning rate and momentum make a huge difference.

IRprop- is the most robust variation of the RPROP algorithm. In the following we provide the pseudo code for the update of a gradient with IRprop-:

```
# Input:
# - grad: the gradient of the error function
# - parameter: the old values of the free parameters before the update

change = grad * last_grad
if change > 0:
    delta = min(delta * pos_step, max_step)
if change < 0:
    delta = max(delta * neg_step, min_step)

parameter = parameter - sign(grad) * delta

if change >= 0:
    last_grad = grad
else
    last_grad = 0

# Output:
- parameter: the new values for the free parameters
```

2.3.4 Regularization

There is more to training a model than only optimization. As the goal of machine learning is to perform well on unseen observations, it is necessary to achieve a good *generalization* when learning the trainings examples. A model with enough parameters could learn the data by heart, which would mean that the model does not understand the data and will perform very bad on new observations. This phenomenon is called *overfitting* the training examples and there are several *regularization* techniques that counter this effect.

The main reason for overfitting are irregularities in the data set. Some of these irregularities have an actual correlation with the input to output mapping, others are just noise that is caused by a sampling error of the training set. It is impossible to tell which of these irregularities is real and which ones are accidental. The model fits both kinds of these irregularities and if the model is flexible, it will fit the accidental irregularities perfectly and have a bad generalization.

The best way to prevent overfitting is to simply get more data. If you can get more data, there is no reason to try to approximate the noise or limit the networks capabilities. More data is always better as long as it is computationally feasible. However, if the items in the training set are correlated and distinct from the examples in the test set, then even more training data cannot prevent overfitting.

Another way to ensure a good generalization is to limit the capacity of the network. This is done by leaving the model just enough flexibility to fit the true irregularities well, but not enough to fit the accidental ones. Of course, this is very difficult to do and generally assumes that the irregularities of the noise are smaller than the meaningful ones.

2.3.4.1 L_1 and L_2 Regularization

L_1 and L_2 regularization is a way to limit the networks capacity and works by adding an additional term to the loss function that penalizes extreme parameter values:

$$E(\theta, D) = NLL(\theta, D) + \lambda \|\theta\|_p^p$$

$$\|\theta\|_p^p = \left(\sum_{w \in \theta} |w|^p \right)^{\frac{1}{p}}$$

which is the L_p-norm. λ is a hyper-parameter that controls the influence of the regularization term on the error function. p is usually 1 or 2.

The effect of this regularization term is that the parameter values are more evenly distributed. Large parameter values are punished which limits the degree of nonlinearity that can be modeled. This greatly prevents overfitting and favors a simple solution that fits the trainings data well, which usually means that a good generalization is achieved.

It was shown that L_1 performs better than L_2 if there are a lot less trainings examples than features, because L_1 performs a sort of feature selection on the data set [76]. On the other hand, if all features are very meaningful, then L_2 is expected to perform better as it just smoothes the network mapping without removing features. Since the use of L_1/L_2 is very case dependent, it is common to use the sum of both L_1 and L_2 regularization with respective hyper-parameters that can be tuned. The exact values of the coefficients for the regularization are often determined using cross-validation and empirical testing.

2.3.4.2 Early stopping

Early stopping prevents overfitting by monitoring the learning process of the network. The technique makes use of an additional validation set where performance of the model is tested on new and unseen observations. If there is no more improvement, or worse, the performance on the test set degrades, then the learning process is aborted. The validation set is split off from the training set and is independent from the test set. Therefore, the network is allowed to test its performance on the validation set at each epoch.

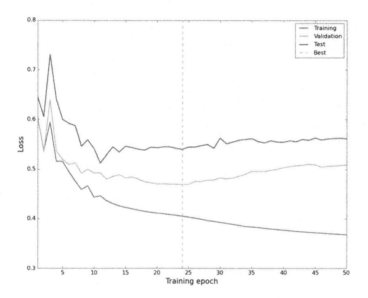

Figure 2-9: Example of early stopping to prevent overfitting [54]

During the trainings process of a neural network (see Figure 2-9), the error on the training set usually decreases continually, while the errors on the test and validation set increase at some point. At this point, the model starts to adapt too much to the trainings examples, which causes a loss in *generalization* and a decrease in performance for new observations.

2.3.4.3 Input noise

Adding Gaussian noise to input vectors is a way of increasing the size of the training set. This is a very good way to prevent overfitting and increase the generalization, especially when small training set is small [77,78]. Using this technique usually results in a worse performance on the training set, but a better performance on the validation and test set.

The amount of noise to add highly depends on the data set, but the new observations should be very similar to the original ones. If the noise is too big, then the model will be unable to learn the data. Experiments have shown that adding half the variance is usually a good start.

2.3.4.4 Dropout

The idea of dropout [79] is to get a large number of neural networks, each trained separately by leaving out parts of the trainings data. Similar to the concept of a random forest, a dropout network is an ensemble of many classifiers that form a consensus prediction when evaluating. As training one neural network is quite expensive, separately training many would not be feasible.

Dropout is a way of training one very large model that combines many neural networks, without actually having to train them separately. For each training step, half of the hidden units are randomly deactivated. This results in a different network architecture for each trainings case. However, a hidden unit is not sampled only once, but many times for different architectures. This results in a great amount of weight sharing for the different models and it is the key idea behind dropout.

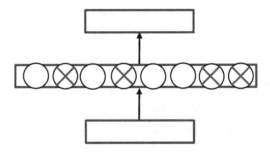

Figure 2-10: Example of dropout with one hidden layer [54]
In this example, four of nine hidden units are randomly deactivated by sampling with a probability of 0.5. The network is trained only using the active units, resulting in the weights having twice their normal value on average.

Consider the example in Figure 2-10 with a feed-forward net with one hidden layer. A neuron in the hidden layer is omitted with probability 0.5. In this case, the four crossed out nodes were deactivated. When now training the network, these hidden units are ignored and active ones are used. Note that the sampled architecture is only used for one trainings example and a new dropout architecture is sampled. It is evident that 2^8 different architectures can be sampled in our example with eight hidden units.

When using a hidden unit, it has the same weights as in all other sampled architectures. This weight sharing makes the model strongly regularized. It is a much better regularization technique than L1/L2, which just pulls the weights towards zero. Instead, dropout has a regularization effect that pulls the weights towards the correct value.

At test time, we could again sample architectures with the same probability, run the test set hundreds of times and take the mean of the output. This would be very expensive and often prediction needs to be done in real time. A much better solution, and this is the second key idea behind dropout, is to use all hidden units but half all their outgoing weights, meaning each weight is multiplied by 0.5. For a network with one hidden layer, this is exactly equal to computing the geometric mean of the predictions of all possible architectures.

For multiple hidden layers with dropout, the weights can also be halved. This is not exactly the same as computing the mean from all different architectures, but it is close enough and very fast. Further, dropout can also employed in the input layer. Usually with a lower probability of dropping an input than a hidden unit, because the input might be vital for the correct output.

3 Using RNNs to predict the lineage choice of stem cells

3.1 Methods

3.1.1 Biological data

The time-lapse movies used in this thesis were conducted at the Institut für Stammzellforschung (ISF, Helmholtz Zentrum München) in the group of Dr. Timm Schroeder.

3.1.1.1 Experimental setup / Data acquisition

A healthy mouse strain that shows no phonotypical difference to a wild type mouse was used as a model organism. From the extracted bone marrow of the mice, HSCs were isolated by fluorescence activated cell sorting (FACS). HSCs were identified by antibodies that bind specifically at the CD150 surface proteins, which allows for specific targeting of HSCs and their descendants. The extracted HSCs were put on a plastic slide that enables cell growth over a long period of time.

During the growth, the cells were observed using an inverse fluorescence microscope namely AxioVert200 (Zeiss) and repeated images were taken by an AxioCAM HRM (Zeiss) camera. Since the used camera was not able to cover the full slide in a single image, the plate was split into an overlapping grid of sub images, called positions. To cover all the positions, the camera was attached to a motor that automatically moves the camera in a programmed pattern. As a result, bright field images were taken in an interval of usually 90 seconds and fluorescence images in an interval of 22.5 minutes.

After performing the experiments, the brightness differences in the bright field images were normalized [80]. The cells in the images were manually tracked over time with Timm's Tracking Tool [46]. This resulted in cell genealogies as discussed in chapter 1.3.

In addition to the bright field images, fluorescence measurements were also taken at certain intervals. This allows for an annotation of the expression values of the biological markers. The transcription factors PU.1 and GATA, as well as the surface protein FCgamma correlate with the differentiation of MEPs and GMPs [53]. The known correlations with the different cell types are as follows:

- FCgamma rises in GMPs
- GATA rises in MEPs
- PU.1 drops before GATA expression rises
- GATA is not expressed when PU.1 expression rises

Using FCgamma and GATA as markers, MEPs can be distinguished from GMPs very reliably. As of now, PU.1 is not used to label the cells, as its exact expression profile is still unclear during the differentiation of MEPs or GMPs.

The question this thesis addresses is whether MEPs can be distinguished from GMPs using only bright field images that depict the morphology of the cells. This was addressed in the thesis of F. Buggenthin [55] and we want to extend on these results. Further, we want to investigate whether PU.1 can be used as a marker. In the ideal case, one of these methods allows the identification of MEPs and GMPs before GATA and FCgamma are able to.

3.1.1.2 Data used in this thesis

For testing purposes, four time-lapse movies were available, each with bright field images at slightly different time intervals. Each cell cycle is counted as one cell. After the cell divides, we start counting two new cells. During some cell cycles, the biological markers pass a threshold and the cell as well as all its descendants can be labeled accordingly.

Name	Unknown	MEPs	GMPs	Time interval (sec)
120602PH5	705	213	489	153
130218PH8	544	337	1441	98
130708PH9	78	17	34	120
140206PH8	237	39	26	99

Table 3-1: The four time-lapse movies used in this thesis and their cell counts. The 120602PH5 and 130218PH8 movies are the core of the data set, as the other movies contain only few MEPs and GMPs.

When trying to distinguish the cells, we can look at different time windows. From only using a single image, to cell tracks over one cell cycle or even including the cell cycles of the cell's ancestors to make a decision. In preliminary tests, we found that looking at one single image is not enough. This led us to start looking at complete cell cycles and their ancestors where we achieved good results in first tests. However, we shifted our focus to only looking at one cell cycle of a cell for classification, as this led to good results while keeping the method practical. When using the method for live prediction, we have to rely on automatic cell tracking and for tracking algorithms, handling cell divisions is very difficult. Therefore, we focused on tracks of a cell over one cell cycle for practical purposes, but using full lineages should also be pursued (see chapter 4.6.4).

To normalize the time intervals (see Table 3-1), we chose to leave out every second image for 120602PH5 and only take every third image for 130218PH8, 130708PH9 and 140206PH8. Therefore, the time intervals are 300 seconds in all movies except 130708PH9 with 360 seconds. However, since this movie has only a small amount of tracked cells, this difference is unlikely to affect the results. The fluorescence measurements were interpolated with the nearest neighbor method for every time step where a bright field image is available.

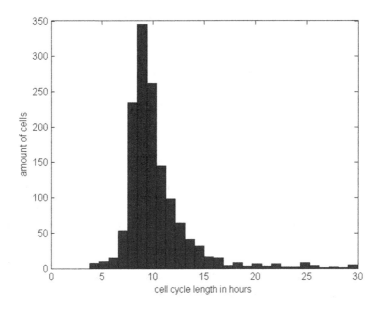

Figure 3-1: Histogram of cell cycle length of the 120602PH5 movie [54]

The lengths of the cell cycles vary (see Figure 3-1). After time step normalization, the cycles have an average length of 130 time points, which means that each cell cycle is roughly 11 hours long. Note that we only normalize the time interval between two images such that experiments with different settings become comparable; the cell cycles are not length normalized.

3.1.1.3 Training and test sets

We aim to build a classifier that can understand time, learn from the cell tracks on one movie and apply this learnt knowledge to predict the labels of cell tracks of another movie. An easier task would be to perform the prediction on other cells of the same movie the classifier was trained on, since the classifier could learn the movie dependent biases. However, this is not practical, as it should be possible to apply the method during live prediction for a new movie.

As we only have two movies with large amounts of tracked cells (see Table 3-1) available, we create two different pairs of training and test sets as follows:

(A)   Training: 130218PH8, 130708PH9, 140206PH8
                    859 unlabeled, 1932 labeled cells
       Test: 120602PH5
                    705 unlabeled, 702 labeled cells

(B)   Training: 120602PH5, 130708PH9, 140206PH8
                    1020 unlabeled, 818 labeled cells
       Test: 130218PH8
                    544 unlabeled, 1778 labeled cells

Merging the movies results in a more general data set. As the test sets are distinct from the training set, having different movies in the training set is vital to achieve a movie independent generalization. The disadvantage when adding movies to the training set is a longer training time. Other than that, merging movies should not be a problem and only result in a better performance if the data is correct. A concern is that the data sets are imbalanced with 70% GMP in training and 80% in the test set of data set (A). The imbalance for (B) is 80% GMP in the training and 70% in the test set. However, since the training and test sets have roughly the same imbalance, adding an additional normalization prior is not necessary.

As an additional validation set is required for early stopping when training neural networks, a random 15% of the observations in the training set were split off and used to measure the networks performance on unseen data.

To evaluate our method, we mainly look at the performance when learning on labeled cells of the training set and predicting labeled cells of the test set. However, to investigate whether the cells can be distinguished before the marker onset, we also need to perform our prediction on the unlabeled cells. In order to obtain a label for the unlabeled cells, all the descendants of a cell in the tree were checked whether their type in the in both descendants agrees, meaning only Unknown and MEP or only Unknown and GMP. If this was the case, the cell and all its descendants were labeled accordingly (see Figure 3-2).

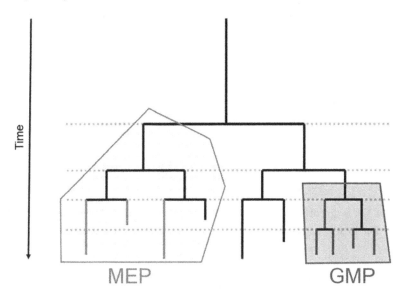

Figure 3-2: Backpropagation of labels [54]
The sub-trees of cells where both descendants have the same label, are all labeled accordingly when back-propagating the labels.

In the data set, only cells with label MEP or GMP are used. The remaining cells with class Unknown, where no label could be propagated, are not added to the data set, as it is unclear which label they will develop. In very rare cases, a cell is labeled as MEP and GMP at the same time. This is the case, when the activations of both biological markers become active. These cases are mislabeled and excluded from the sets. When looking at the tracking trees, we often observe that the label of all descendants agrees and most cells can be labeled accordingly. If the label of the descendants does not agree, then the cell is not added to the data set.

3.1.2 Extracting features from images

As a result of the tracking, the position of each cell in the bright field images is annotated. Previous work in the ICB focused on segmenting the cells in the images from the background and extracting meaningful features from the segmentations [81]. Using the maximally stable extremal regions (MSER) algorithm, the cells can be segmented and a bounding box around the cell can be established [82]. The result is a sequence of small cell images for each cell over its cell cycle (see Figure 3-3).

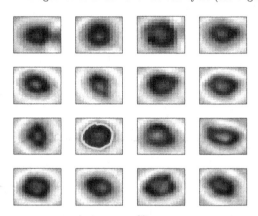

Figure 3-3: Segmented bright field images of an exemplary cell over one cycle [54]
The highlighted boundary indicates the manual segmentation of a cell.

Based on this segmentation, features can be derived that describe the cell's shape and texture. The diploma thesis of F. Buggenthin [55] combined several methods to calculate 88 features for each bright field segmentation of a cell:

- Shape: area, axis lengths, eccentricity, orientation, perimeter, diameter
- Colors: intensity, contrast, correlation, variance
- Histogram of Oriented Gradients features [83]
- Histogram of Zernike moments [84]: similarity to predefined shapes
- Tamura features [85]: coarseness, contrast, directionality, linelikeness, regularity, roughness

In addition to the features from the bright field images, the movement speed and the intensity of PU.1 are also measured. However, we first focus on bright field images and only use the PU.1 intensity as an alternative case as it requires transgenic mouse strains that express fluorescently labeled PU.1 proteins.

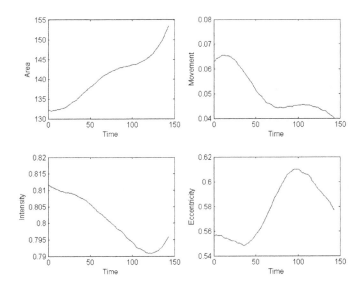

Figure 3-4: Exemplary features of a cell over one cycle with smoothing [54]. The area steadily increases while the movement declines as the cell grows. During the growth, the cell becomes less oval-shaped until it divides again.

When looking at the features of cells over one cycle (see Figure 3-4), we repeatedly observe the same behavior. During a cycle, the cell grows and increases in area size. Further the eccentricity of a cell increases, meaning it gets less oval-shaped, during the middle of the cycle and then declines again. We hypothesize that MEP and GMP show differences in these features during the course of one cell cycle and that these differences can be used to distinguish the cells.

3.1.3 Baseline with SVM

To test whether it is at all possible to distinguish MEPs from GMPs using the time series of the extracted features over one cell cycle, we perform a simple test using a Support Vector Machine (SVM) for classification [15]. In its nature, a SVM cannot understand time and can only handle a fixed amount of features.

Therefore, each cell cycle is first normalized to the same length and the time series of a feature is described with only a few points. Instead of one measurement for each time point, an estimate of the curve using 10 evenly spaced values is obtained. A more complicated alternative is to fit a spline as used in the prior approach [55], but this did not improve the results. Note that both approaches normalize all cell cycles to an equal length, which is very questionable. For instance, if MEPs would simply have longer cell cycles than GMPs, then this approach would fail to detect that. Since the SVM cannot handle time, the time dimension is collapsed and the time points of a feature are treated as independent observations. In our case of 88 features with 10 normalized time points, this results in a feature vector of length 880.

For each time series in our training set, the SVM classifier is trained on the given label for the observation. Afterwards a prediction for every cell in the test set is made, resulting in a probability value whether the cell is MEP or GMP. To evaluate the performance on the test set, the Area Under the Curve (AUC) is used as a measurement of correctness. For a small data set, the AUC is much more reliable than merely using the accuracy, as the accuracy only differentiates between correct and false. The AUC however also takes the certainty of the prediction into account, which is expressed in the probability value.

3.1.4 Recurrent Neural Networks

As a model that can understand time, our method of choice are the RNNs. Deep feed forward neural networks have achieved great success in recent years, but its recurrent counterpart was still lacking to show good results. However, just a few months before this thesis and additionally during the writing of this thesis, two very promising new approaches were developed. One using classic RNNs with dropout [86], the other using LSTM cells in a multi layered and bidirectional model [67]. We focus on these two approaches and build models on top of this work.

In addition to the training and test set, a validation set is also used when training neural networks to allow for early stopping. With a validation set, the performance on a set similar to the training set can be observed. This shows when the decrease in error on the training set is not due to a better general understanding of the data, but due to overfitting. For every trainings procedure, we observe the performance on the validation set and the parameter combination that gave the best results on this set is the result of the trainings process.

3.1.4.1 Dropout RNN

The basic RNN is very hard to train as it suffers from the vanishing gradient problem we described in chapter 2.2.2. This issue has recently been tackled by the introduction of the hessian-free optimization technique [74], which is very sensitive to small gradients by approximating the second derivative of the loss function. However, using this optimization technique and testing different network architectures is hardly feasible, as one optimization step takes several hours even for a small data set.

Recently it was reported that the success of training neural networks greatly depends on the initial parameters of the network [71]. When using the initial values described in chapter 2.3.2 for the RNN, the network can be trained even without using the hessian-free optimization. Since the training set is small with about 2000 observations, using a full batch learning is optimal. We found that using RProp as described in chapter 2.3.3.2 is an excellent general-purpose optimization method, which produces great results on all tested architectures and bypasses the vanishing gradient problem by only considering the sign of the gradient.

During the writing of this thesis, an extension to RNNs [86] by using dropout (see chapter 2.3.4.4) was published. Specifically, the authors are using an approximation to dropout called fast dropout [87]. As our network was overfitting, we implemented this architecture. We use dropout on the input-to-hidden connections and the hidden-to-hidden connections (see Figure 3-5) as described in the paper. A concern was raised that using dropout on the hidden-to-hidden connections messes with the ability of the recurrent network to model sequences [88]. In our case however, using dropout only on the input-to-hidden connections was not enough to prevent overfitting.

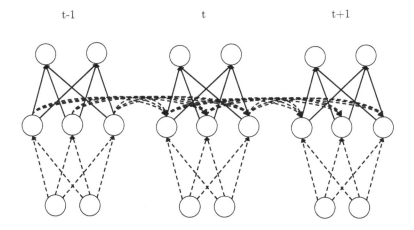

Figure 3-5: RNN with dropout unfolded in time [54]
Dropout is used on the input-to-hidden and hidden-to-hidden connections as indicated by the dashed lines.

We compared the performance of the classic RNN to the implementation with fast dropout to see the impact of the regularization on the network.

3.1.4.2 Bidirectional Deep LSTM

As a second approach, we use a recurrent network with LSTM units as described in chapter 2.2.3. By doing so we avoid the problem of the vanishing gradient as the LSTM can naturally model very long time sequences with its memory cells. This even allows us to train the network with SGD. However, we opted to use RProp to avoid any hyper-parameters and we even observed faster convergence with RProp. All the weights of the LSTM network can be initialized with mean zero and a Gaussian standard deviation of 0.01.

It was reported that extending a basic LSTM model with a bidirectional approach helped when performing frame wise classification [89]. In our case, the exact timing of the predictions is not important and for sequence classification, the prediction only needs to be right on average per sequence. However, we observed that the output of the RNN at the first few time steps of the RNN is often useless, as the net has only seen a few time points at this stage. Therefore, we also employ the bidirectional approach for sequence classification, which enables the network to make proper predictions at every time step as it has always seen the whole sequence.

A further extension to the bidirectional LSTM model was published very recently [67] which uses a hybrid model with stacked bidirectional networks in a layered fashion. This results in a network that is not only deep in time, but also deep in space. The idea is very simple: we copy the hidden layer of a bidirectional LSTM and put it on top as a second or even third layer (see Figure 3-6). It was shown that this increases the performance and the level of abstraction that the recurrent network can achieve. We get a boost in performance for the first additional layers, but after adding three layers there is no more gain.

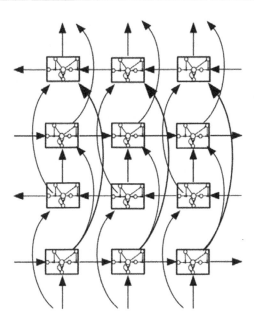

Figure 3-6: Deep bidirectional LSTM with two recurrent layers each, adapted from [67]. The concept is similar to a bidirectional RNN, except that LSTM units are used and that hidden layers are stacked on top of each other. However, the stacking of the forward and backward direction is independent, and the information is only merged at the output layer.

We are interested in the performance of the normal one layered forward LSTM in comparison to the deep bidirectional LSTM (DB LSTM).

3.1.4.3 Output Layer

A RNN gives an output for every time step. To perform sequence classification, these outputs have be merged together into one output for each sequence. The general approach is to sum up the values of all the output units over the different time steps and then push them either through a sigmoid activation function in the case of binary classification, or into the softmax function when having multiple classes. The act of combining the values from different time points, or more generally from different layers, is called pooling.

In our experiments, we observed that first pooling the output and then applying the sigmoid function makes the network very hard to learn. The reason is probably that optimizing this function while not overfitting is very hard for long sequences, as one time point can already have a huge impact on the classification. Instead, we achieve better results when first applying the sigmoid function to the output at every time step and then pooling the time steps by calculating the average of the sigmoids. In this case, each output is not weighted as heavily and the prediction relies more on being correct on average. The result is a value between zero and one that can be used as a binary probability output for the network.

We also tried to use different methods of pooling for the sigmoid values, like weighting the different outputs by how much of the sequence the model has seen so far in the case of the unidirectional RNNs. For example, a linear weight favoring the later outputs could be used, but this did not improve our results. Also just using the last output of the network might seem like a good idea at first because only at this point has the unidirectional network seen the complete sequence, but this approach is very prone to overfitting.

3.1.4.4 Implementation details and hyper parameters

Training deep neural networks is very time intensive. Therefore, we implemented all methods from scratch to allow for optimized GPU calculation. For floating point operations, graphic cards often outperform CPUs by a large margin, sometimes with a 20-fold increase in speed when running code on the GPU. In our case we use Theano [90], a framework that directly generates GPU optimized C code from mathematical expressions.

To maximize the efficiency of the GPU calculations, our method operates on a 3-dimensional tensor instead of 2-dimensional matrices. Usually, the RNN iterates over the rows of a matrix, where each matrix contains the data of one training example with a row for each time point and a column for each feature. Therefore, only one vector is calculated at a time when applying the equations. However, since the data set is small and we only want to use full batch training methods, all calculations of one time step can be performed at once for all sequences. The input is structured in a 3d-tensor with the first dimension being the time points, the second one the sequences and the third one the features. The loop iterates over the time points and performs the equations in one step for the matrix that comprises all sequences with their features. After looping when all calculations are finished, the first dimension is swapped with the second one to put everything in order.

During our tests, we found that using 50 hidden units per layer for all four architectures is ideal in our case. This is surprising, as one would expect that with e.g. a dropout rate of 0.5 we would be able to double the amount of hidden nodes in the network. A reason for this is probably that fast dropout does not require additional hidden units to achieve the same improvement. We use two recurrent layers for the DB LSTM network, as we found that our results did not improve when using more than two recurrent layers. For all other architectures only one recurrent layer is used.

Each model is trained with Rprop- since the training set is small and we want to avoid the vanishing gradient problem of the long sequences. The weights of the LSTM networks are initialized with mean zero and a standard deviation of 0.01. For the RNN like architectures, we use the parameter initialization protocol described in chapter 2.3.2.

In summary, we will present detailed results for four different types of RNNs. First, the basic one layered forward RNN as well as the extension with dropout. Third, a one layered LSTM network and also a two layered LSTM network with bidirectional connections.

3.2 Results

3.2.1 RNNs based on morphology yield high performance on labeled data

We compared the different recurrent architectures by testing their performance on both data sets (A) and (B) as discussed in chapter 3.1.1. In order to make the sets as clean as possible, the models were only trained and tested on labeled data in this comparison. As the main goal of this thesis is to build a RNN that can predict the label of the cell by just looking at the bright field images, only the morphology features and no additional PU.1 expression levels were used in this part.

The preliminary baseline approach with SVM yielded quite good results with an AUC of 0.79 on both test sets (see Table 3-2). However, the features had to be normalized by their mean and standard deviation of each movie separately, as not doing so lowers the AUC by about 10%. This result indicates that predicting the label from observing the cell's morphology over one cell cycle is very promising. We further compared the performance of the different recurrent architectures on both data sets.

Classifier	AUC A	AUC B	Mean AUC
SVM	0.789	0.786	0.788
RNN	0.776	0.891	0.833
Dropout RNN	0.832	0.884	0.858
LSTM	0.837	0.902	0.870
DB LSTM	0.866	0.935	0.901

Table 3-2: Results of classifiers on unseen test sets from different movies. The RNN approaches outperform the SVM approach significantly, while the best results are achieved by the DB LSTM model. There is a difference in performance when comparing set A and B since the movies are different. Overfitting was prevented using early stopping on a validation set which was split off from the training set. Therefore, the validation set remains independent from the test set. The classification accuracy is around 3-5% higher for GMPs with every model.

We notice that our performance on the test sets improves quite a lot when using RNNs compared to an SVM. However, the results vary slightly for the two different data sets. While the SVM has a stable performance, the results for the recurrent models are more diverse. The basic RNN performs much better on set B than set A. When introducing dropout, the result becomes more stable. It is interesting to note that the performance on set B decreases marginally with dropout.

Looking at the results for the LSTM architectures, we observe a noticeable improvement compared to the classic RNN models. This means that the model benefits from the additional memory cells. Even the unidirectional one layered LSTM achieves better results than the RNN with dropout. The accuracy can be increased further by using the DB LSTM architecture with the proposed bidirectional deep layers. The DB LSTM provides the best performance on both data sets and is therefore our model of choice for any further experiments.

To understand the difference in performance of the recurrent networks on one data set, we look at the trainings process for each architecture in more detail (see Figure 3-7). The AUC is measured during the training process during the first 50 epochs. For each epoch, we measure the performance on the trainings, validation and test set. Note that the trainings and validation set are both from the same set of movies, while the test set is from a different time-lapse movie. For each of the different architectures, we observe that the networks start to overfit on the training data after about 15-20 trainings epochs. The optimal performance on the validation set is marked with the dashed line.

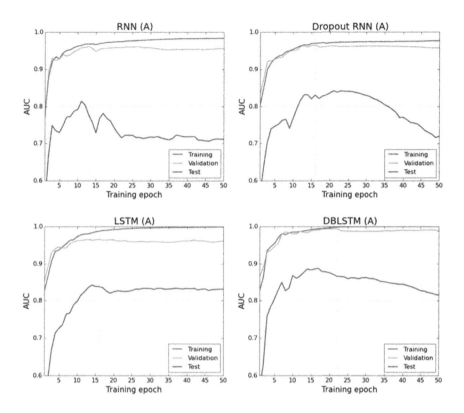

Figure 3-7: Comparison of the classifier performance on set A [54].
The dashed line indicates the best result on the validation set using early stopping.
The best performance is achieved by the DB LSTM model. There is a gap between
the AUC on the validation and test set, since the test set contains cells from a
different movie.
(figure labels are ordered top to bottom)

For the basic RNN and LSTM we can see a noticeable disparity in performance between
the training and validation set. This means, the model overfits heavily on the training
set and does not learn a general structure of the data. In the case of RNNs, this problem
can be addressed by using dropout. Using dropout, we observe that the performance
on the training set degrades, while the performance on the validation set increases.
This means the model has a better generalization, which also results in a better
performance on the test set. For the LSTM models, we see that the model is able to fit

the data better in general and has almost perfect results on the validation set. Unlike the simple unidirectional LSTM, the deep-layered bidirectional LSTM is able to transfer this into a high AUC value on the validation set and achieve a near perfect performance with an AUC of 0.995 on the validation set.

A concerning issue however is the discrepancy between the performance on the validation and test set for every architecture. The best result on the test set is achieved by the DB LSTM, which also performs best on the validation set, but the difference is substantial, especially for set A. When looking at the accuracy on the validation set, it looks like the DB LSTM has done its job perfectly, given the set of features it has available. The reason for the difference in performance is due to the training and test set being very different which will be discussed in chapter 4.

3.2.2 RNNs have long training time but short test time

The RNN has a long runtime for the training process, even when running the code on the GPU. Depending on the size of the training set, several hours is common (Table 3-3). This means that the amount of training examples is very limited and a training set above 10000 examples is problematic.

Model	Training time in minutes
RNN	127
Dropout RNN	163
LSTM	189
DB LSTM	396

Table 3-3: Training time on GeForce GTX 570 of different RNN models
using 3000 sequences of length 1000 in the training set and training for 100 epochs

The recall time of the RNNs however is comparably small, ranging from 20 to 80 milliseconds per example. This means that the time for classification is significantly lower than the runtime of the movie and the method can be used for live prediction.

3.2.3 RNNs can predict lineage choice before biological marker

A practical application is to apply our classifier to unlabeled cells and possibly predict the label of the cells before the biological markers. To evaluate this, we also add all the unlabeled cells to our test set. For each cell, its descendants are checked and if it has only MEPs or GMPs descendants in the tree, the cell is labeled accordingly.

In our simulation, we use the DB LSTM model, which was the best classifier in the last chapter and apply it to the full test set with all cells. It is clear that there is a high correlation between experiment time and classification accuracy. The later in the experiment, the more cells will be differentiated and reveal their label. To see if the classifier can identify the cells before the biological marker, we plot the classification AUC of the cells over experiment time as shown in Figure 3-8. Besides only using the morphology, the network is also trained with the expression profile of PU.1 to test whether PU.1 is viable as a marker. In this case, only PU.1 was used as a feature, as also adding the morphology did not improve the results.

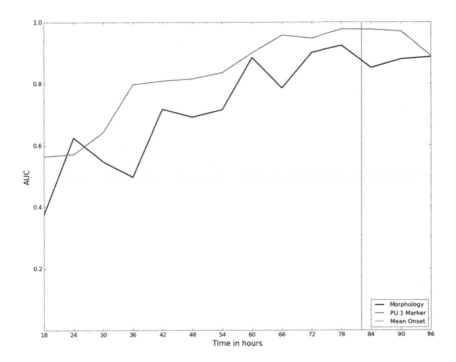

Figure 3-8: Performance over time with morphology or PU.1 on set B [54].
The prediction accuracy increases as the experiment progresses and the cells differentiate. The average onset of the label is at 83 hours with a standard deviation of 13 hours. At this point, around 90% of the cells can be classified correctly using the morphology or PU.1 approach.

The average onset of the GATA and FCgamma markers is at 83 hours, with a standard deviation of 13 hours. In both cases, for morphology and PU.1, the accuracy increases with experiment time. The closer the measured cells are to the onset of the biological markers, the better the classification result. Using only the morphology, the classifier reaches a high AUC of over 0.8 after about 60 hours into the experiment, which is more than 20 hours before the biological markers become active. With PU.1, the network is even able to get a decent AUC of 0.8 after 40 hours in to the experiment.

3.2.4 Accuracy can be improved by only taking high-confidence predictions

In order to evaluate the applicability of our classifier in practice, we extend the simulation from the last chapter and focus only on unlabeled cells, which have no active biological marker. We monitor the amount of cells over time and perform the classification of cells on an unseen movie (set B).

The experiment time is divided into intervals of 6 hours. For each interval, the total number of different cells is counted. Each of these cells is classified using the DB LSTM model. To keep the precision as high as possible, only cells where the classifier is very confident with a prediction score of over 0.9 for one class are taken. As we are interested in the performance before the biological marker onset, only cells are counted that have not been labeled.

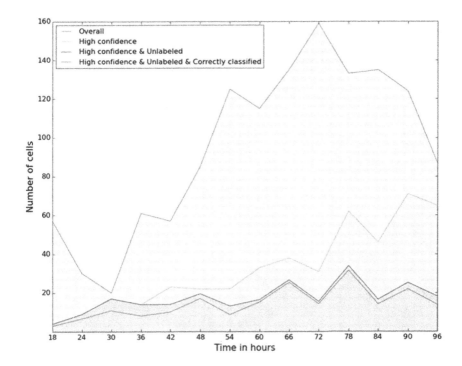

Figure 3-9: Performance over time for unlabeled cells using morphology [54]. At 60 hours the classifier achieves a high accuracy on the unlabeled cells (proportion of green to blue area), when only using high-confidence predictions.
(figure labels are ordered top to bottom)

For the morphology (see Figure 3-9), the classifier has a high accuracy of over 80% after 48 hours and a very high accuracy of over 90% after 60 hours when only using high-confidence predictions. It is interesting to note, that the overall amount of cells grows faster than the amount of cells where the classifier predicts with a high confidence. Still, more than 90% of the high confident cells are unlabeled after 48 hours, and more than 60% are unlabeled after 60 hours. The time until the onset of the cells between 48 and 60 hours is 21 hours on average. This means that the classifier is clearly able to predict the cells before the marker onset by only using the morphology.

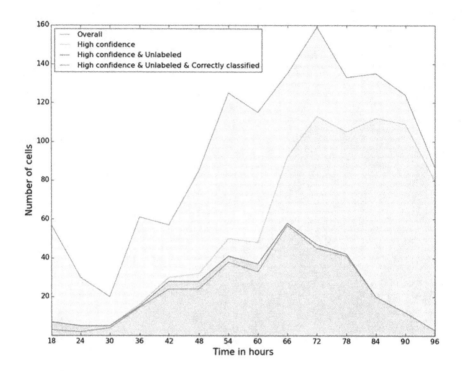

Figure 3-10: Performance over time for unlabeled cells using only PU.1 [54]. At 60 hours the classifier achieves a perfect accuracy on the unlabeled cells when only using high-confidence predictions.
(figure labels are ordered top to bottom)

For the classifier using only PU.1 (see Figure 3-10), the results are different. The classifier is generally more confident and the amount of high confident cells is proportional to the overall amount of cells. Interestingly, the accuracy at 42 to 60 hours is not better than the one for the morphology. Nevertheless, at around 30 to 36 hours and after 66 hours, the PU.1 classifier has a perfect accuracy of over 95% and performs better than the morphology method. Note that all accuracies were calculated using a different movie in the test set (set B).

3.2.5 Shape and Texture are important features

To analyze which of the features are key for the machine learning models to understand the data set, we perform a feature reduction. This means that parts of the features are left out from the training and test process, and the performance only using the rest of the features is measured. In chapter 3.1.2, we already discussed the different types of features available. As leaving only a few features out made no observable difference and training a network takes several hours (see chapter 3.2.2), we decided to divide the features into groups and only use one of the groups per test case.

Feature set	AUC
Movement, Shape	0.849
Zernike moments	0.701
Entropy, Gabor	0.876
Tamura	0.891
HOG	0.537
All features	0.902

Table 3-4: Feature importance using DB LSTM on set B.
The basic features like shape and texture are very important. More advanced features like Zernike polynomials are less informative.

Our results for this analysis (see Table 3-4) are very informative. We found that movement and shape are very good indicators for the class of a cell. Even more important seems to be the surface of the cell that is described by the Tamura features. The other features describing the entropy and variance also seem to be very important. As the features were calculated on the segmented cell images without a fixed image size, the Histogram of Oriented Gradients (HOG) features are not informational.

4 Discussion

The results with the baseline SVM approach show that distinguishing between MEP and GMP using the morphology of the cells is possible. We achieve a good accuracy even though the time series is averaged for each feature to only a few time steps. However, the features had to be normalized by their mean and standard deviation of each movie separately, as not doing so lowers the AUC by about 10%. This procedure makes the features more robust.

4.1 Different performance on movies

In the results, we noticed a better performance on set B than A. In both cases, the majority of the training and test set each consist of one large movie with many cells. When swapping the movies, one might expect the same performance on both test sets, but there are a couple of different scenarios, which can cause the performance to degrade.

The first explanation that comes to mind is that the learning process is worse for set A which results in a lower generalization. If the training set contains many false positives or cells which look slightly different than the average MEP or GMP, then the model could not learn the data properly, which would result in a lower accuracy on the test set. However, in this scenario, we would also expect a bad performance when swapping the training and test set.

A more likely explanation is that the 120602PH5 movie, which is the test set in dataset A, contains some observations that are very distinct from the other movies. For instance, if the cells in this experiment were in a significantly later stage of differentiation due to some external factors in the experiment, then the MEPs would be larger in the movie. Another scenario is that this movie contains more cells and is more crowded in the later stages than the other movies, which would cause the segmentation to fail. In both cases, these cells could not be classified in the test set. However, as there are still enough proper observations we would not notice this problem when the 120602PH5 movie acts as the training set.

4.2 Robustness of features

The more complex recurrent model is able to give a much better performance on both sets. Nevertheless, there is still have a noticeable gap between the validation and the test set. The optimal performance on the validation set suggests that the features are well chosen. However, the accurate performance could also stem from biases in the movies.

Each movie generally has the same conditions and some of them were even conducted using the same microscope, but when looking at the bright field images, we often notice a big difference in brightness. Even when comparing different position in one movie, there is a noticeable change in brightness. In a post-processing step, the brightness across positions is normalized by estimating the luminosity and contrast of the different positions and then applying filters to equalize the difference but this is still far from optimal.

In our preliminary tests, we observed a perfect accuracy when classifying on the same movie within a cross-validation framework. Later we found, that the good performance was due to the movie being darker on the right side and the fact that MEPs were located more on the right, while GMPs were more on the left plate of the movie. Therefore, our machine-learning algorithm was able to distinguish the cells by merely looking at the brightness level. From this point forward, we completely abandoned the idea of using the same movie in the training and test sets and focused on classifying movies that are distinct from the training set.

In an ideal setting, the features would be independent from the experimental settings of the time-lapse movies. As the shape and surface features performed best in our feature reduction, we suggest building robust features on top of these.

4.3 Building better features

An idea we pursued was to normalize every cell image by its mean and standard deviation after calculating its bounding box and before performing the segmentation and the feature analysis. This causes the brightness and intensity of the cell image to be dependent on the size and content of the cell image rather than the experimental settings. Different images across one cell cycle would vary greatly in their intensity depending on the size of the cell, but given enough examples, a machine-learning algorithm could understand this and have a set of movie independent features. This procedure also makes normalizing the final features by the mean and standard deviation of each movie unnecessary.

Further, instead of calculating features on a variable size of the images, we changed our approach to image patches with a fixed size of 30x30. This resulted in a better performance when using position dependent features like the HOG. We chose a size of

30, as plotting the histogram for cells before the marker onset revealed that 95% of the cells before the marker onset would not exceed this size. After the onset of the marker, the cells tend to be bigger, but these cells are mostly fully differentiated MEPs and easy to detect using the size criteria alone.

Applying these normalizations resulted in a more closely matched AUC for the test and validation set, but unfortunately, the overall AUC decreased, meaning the performance degraded. We suspect that there was a loss in information when performing the brightness normalization. Interestingly, we were able to increase the AUC significantly when again normalizing the final features by the mean and standard deviation of each movie. Still, the AUC was 5-8% percent lower compared to the approach as reported in our results. This means that despite the normalization, the features are still movie dependent, which requires further investigation.

A different approach we pursued was building completely new features using a Stacked Autoencoder [23]. We did not include this in our methods or results, as this approach yielded no success. The features were overfitting on the movies resulting in a good performance on the validation set, but poor on the test set. Distinguishing MEPs and GMPs is very difficult for the human eye and usually only possible in later stages of an experiment when the MEPs are significantly larger than the GMPs. Automatically creating the features using an unsupervised approach would therefore be ideal. However, doing so usually means that the feature do not describe distinct properties of biological cells, but rather differences between whole image patches and thus different movies. This would mean that all the features are heavily movie biased and therefore an unsupervised approach is unlikely to work until the movie bias is completely removed.

4.4 Limitations of RNNs

Looking at the results, the advantages of using RNNs are clear. The performance of the RNN, compared to other classifiers that cannot understand time, is substantial. However, using this type of model also has its disadvantages. The RNN has a long runtime for the training process, even when running the code on the GPU (see chapter 3.2.2).

Further, the amount of input features has to be below 100 in order for the network to learn the data. This is evident when trying to combine the RNN with other layers, like prepending a convolutional layer before the recurrent layer, which does not work. The reason for this is the vanishing gradient problem that has been discussed. Introducing a full batch learning method like RProp helps, but at the cost of computational time. Using the LSTM neurons, also avoids the vanishing gradient problem. However, the LSTM model needs more parameters and equations, resulting in a longer optimization time.

4.5 High accuracy before biological marker

In our results, we achieve a high accuracy at around 60 hours when using the morphological features and 40 hours when using the PU.1 expression level. This suggests that the lineage decision of the HSC whether to become MEP or GMP is already much earlier than the change in GATA and FCgamma levels.

The disadvantage of the PU.1 approach is that special mice are required which have the fluorescent knock-in sequence for PU.1. This is a very complicated process and requires a long period of preparation. Unfortunately not using this marker and only using the morphological features only gives us a high accuracy at about 60 hours. A tradeoff would be to perform a semi-automatic approach where the tracking algorithm creates small periods of tracks where it is very certain and asks a user to make interactive decisions to connect them.

Both results can be improved by only taking cells, where the classifier has a high confidence in the prediction of the label. This approach increases the accuracy from 80% to 90% in both cases, for either morphology or PU.1. As having false negatives is not problematic as long as enough cells are identified as positive and decreasing the amount of false positives is very important, this step is highly practical.

4.6 Outlook

4.6.1 Learning on raw images

Our great results could be improved by developing features that are more robust. The recurrent neural network is able to classify cells perfectly if they stem from the same movie with the same conditions. However, when cells from another movie are classified, the results drop. This means that the features are movie dependent. The issue can be addressed by either gathering more training data in order for the model to learn the difference between movies, or by building a set of more robust and movie independent features.

Instead of spending a lot of time on making new features, one could try to normalize the individual cell images and build a model that can work directly on the raw images. This approach is very difficult with recurrent neural networks, as they already have to spend a lot of power on understanding the temporal component. However, as the cell images are easy to normalize and bring into a central position of the image, this might be possible with deep-layered temporal model like the DB LSTM. This method was successfully used for speech recognition on mel-cepstrum features of audio spectrograms and cell images are only seem slightly more complex when normalized properly. A concern is that recurrent nets are known to have problems with too many features. The cell images would have around 500 pixel features for each image with a square size of 24 and this is much more than the mel-cepstrum with only 24 features.

4.6.2 Finding specific important features by feature reduction

To build a better classifier, it is important to find out which specific features are important when distinguishing the cell types. This is especially important in our case, as the test set is distinct from the training set, and an improvement might only be possible by manually restricting the features to prevent overfitting on the training data.

We thought about making a complete feature reduction, by iteratively leaving out all possible groups of features, but the training time for the network with several hours is too long. To reduce the amount of different combinations that have to be tested, the Apriori algorithm [91] is often used in these reduction based problems. The central idea is that if leaving one feature out increased the error above a threshold, removing additional features will not decrease the error again. This assumes that the network always finds the optimal parameters for the validation set. However, this theorem does not hold in our case, as removing an additional feature could decrease the error as it prevents overfitting.

Nevertheless, as we already identified subgroups of important features in our results, we could perform a complete feature reduction with all combinations based on these results. When estimating the training time at 3 hours with and performing a reduction of 2 subgroups with 5 features each, this would take 7,5 days of training time on the GPU board.

4.6.3 Predicting cell fates live during movie acquisition

A key step in making the classification method practical is to combine it with an auto tracking approach that can track cells live during movie run-time. This would make on-line prediction of cell fates possible. However, auto tracking approaches are still limited in their performance, especially in later stages of the experiments where the images are too crowded to separate the cells properly. Therefore, a tradeoff in experiment time has to be established where the prediction yields a good accuracy and auto tracking is still possible.

If this method were to be put in practice, it would become feasible to extract cells with high prediction confidence at early stages of the movie. Single cell sequencing could be performed on the extracted cells, which could provide new insights into the biological processes that are involved in differentiation.

4.6.4 Using lineages of cells instead of only one cell cycle

During our work, we mainly focused on using the current cycle of a cell and not the complete linage with all cell cycles of the cell's ancestors. This was due to practical reasons, as the auto tracking approach has problems with cell divisions.

However, once these problems are addressed, using complete linages is likely the better option, as more information about the cell usually leads to better results. In prior tests, we already evaluated this and achieved very promising results. The extended length of the sequences of up to 2000 time points did not cause any problems when using the LSTM models. Also concatenating the cell cycles, without explicitly defining the bounds of each cycle in the sequence, did not seem to be an issue for the model.

A problem, that needs to be addressed when learning full lineages, is that some cells share the same ancestors. These common ancestors are the prefix of the sequence in different linages, which results in the first part of the sequences being identical. If a very large tree is added to the training set with many examples, this prefix is learned many times and the network might only use this prefix to identify the linages. To address this issue, the recurrent architecture of the network needs to be changed. Instead of learning different linages, the network would have to learn a complete tree. This can be implemented by indexing the time steps and using specific previous time steps instead of t_{-1}. These models are called recursive neural networks [92].

5 Conclusion

We successfully managed to distinguish MEP from GMP cells by only using morphological features with a very high accuracy of over 80%. Further, our method can detect these cells up to two cell cycles before the currently used biological markers. When only using the morphology of the cells, we achieve a high accuracy after 60 hours of experiment time, while the biological marker detects the cells at about 80 hours on average in our experiments. If the expression level of PU.1 is used for identification, the cell types can be classified with a high accuracy at 40 hours. Both results can be improved to 90% accuracy by only high confidence predictions.

This enables us to combine our method with an auto tracking approach that is viable for the first few cell cycles of a time-lapse movie where the cells are not as great in number and easier track. Note that we measured our performance for the prediction of cells from time-lapse movies that are different from the training set and therefore the approach is directly applicable in a practical method.

As the recurrent neural networks are general machine learning models, our fast GPU implementation can be used as a general-purpose method for other time-series classification problems in bioinformatics. By slightly changing the architecture of the network, the model could also be applied to other problems in cell imaging like cell segmentation and auto tracking.

References

[1] R. Solomonoff, "An Inductive Inference Machine," *IRE Conv. Rec.*, vol. Section on, 1957.

[2] H. Rogers, *Theory of Recursive Functions and Effective Computability.* 1987.

[3] A. L. Samuel, "Some Studies in Machine Learning Using the Game of Checkers," *IBM Journal of Research and Development*, vol. 3. pp. 210–229, 1959.

[4] C. M. Bishop, *Pattern Recognition and Machine Learning*, vol. 4. 2006.

[5] M. Mohri, *Foundations of Machine Learning.* 2012.

[6] I. V. Tetko, D. J. Livingstone, and A. I. Luik, "Neural network studies. 1. Comparison of overfitting and overtraining," *J. Chem. Inf. Comput. Sci.*, vol. 35, pp. 826–833, 1995.

[7] C. P. M. R. Greene D., "Unsupervised learning and clustering," *Lect. Notes Appl. Comput. Mech.*, pp. 51–90, 2008.

[8] H. B. Barlow, "Unsupervised Learning," *Neural Computation*, vol. 1. pp. 295–311, 1989.

[9] T. J. S. Geoffrey E. Hinton, *Unsupervised Learning: Foundations of Neural Computation.* 1999.

[10] V. N. Vapnik, *The Nature of Statistical Learning Theory*, vol. 8. 1995.

[11] R. A. Fisher, "The use of multiple measurements in taxonomic problems," *Ann. Hum. Genet.*, vol. 7, pp. 179–188, 1936.

[12] E. Alpaydin, *Introduction to Machine Learning.* 2012.

[13] S. L. Gortmaker, D. W. Hosmer, and S. Lemeshow, "Applied Logistic Regression.," *Contemporary Sociology*, vol. 23. p. 159, 1994.

[14] V. Bewick, L. Cheek, and J. Ball, "Statistics review 14: Logistic regression.," *Crit. Care*, vol. 9, pp. 112–118, 2005.

[15] C. Cortes and V. Vapnik, "Support-vector networks," *Mach. Learn.*, vol. 20, no. 3, pp. 273–297, Sep. 1995.

[16] Tin Kam Ho, "Random decision forests," in *Proceedings of 3rd International Conference on Document Analysis and Recognition*, 1995, vol. 1, pp. 278–282.

[17] L. Breiman, "Random Forests," *Mach. Learn.*, vol. 45, pp. 5–32, 2001.

[18] I. H. Witten and E. Frank, *Data Mining: Practical Machine Learning Tools and Techniques, Second Edition (Google eBook).* 2005.

[19] I. Guyon and A. Elisseeff, "An introduction to variable and feature selection," *J. Mach. Learn. Res.*, vol. 3, pp. 1157–1182, Mar. 2003.

[20] R. M. Haralick, K. Shanmugam, and I. Dinstein, "Textural Features for Image Classification," *IEEE Trans. Syst. Man. Cybern.*, vol. 3, no. 6, pp. 610–621, Nov. 1973.

[21] O. Chapelle, P. Haffner, and V. N. Vapnik, "Support vector machines for

histogram-based image classification.," *IEEE Trans. Neural Netw.*, vol. 10, no. 5, pp. 1055–64, Jan. 1999.

[22] D. Rumelhart, G. Hinton, and R. Williams, "Learning internal representations by error propagation," 1985.

[23] Y. Bengio, P. Lamblin, D. Popovici, and H. Larochelle, "Greedy Layer-Wise Training of Deep Networks," *Adv. Neural Inf. Process. Syst. 19*, pp. 153–160, 2007.

[24] R. Salakhutdinov and G. Hinton, "Deep Boltzmann Machines," *Artif. Intell.*, vol. 5, no. 2, pp. 448–455, 2009.

[25] G. Hinton, S. Osindero, and Y. Teh, "A fast learning algorithm for deep belief nets," *Neural Comput.*, 2006.

[26] Y. Bengio, A. Courville, and P. Vincent, "Representation learning: A review and new perspectives," no. 1993, pp. 1–34, Jun. 2013.

[27] A. Graves, "Practical Variational Inference for Neural Networks.," *NIPS*, pp. 1–9, 2011.

[28] A. Krizhevsky, I. Sutskever, and G. Hinton, "ImageNet Classification with Deep Convolutional Neural Networks.," *NIPS*, pp. 1–9, 2012.

[29] G. Hinton, L. Deng, D. Yu, G. Dahl, A. Mohamed, N. Jaitly, V. Vanhoucke, P. Nguyen, T. Sainath, and B. Kingsbury, "Deep Neural Networks for Acoustic Modeling in Speech Recognition," *IEEE Signal Process. Mag.*, vol. 2, pp. 1–27, 2012.

[30] G. Cybenko, "Approximation by superpositions of a sigmoidal function," *Math. Control. signals Syst.*, pp. 303–314, 1989.

[31] D. E. Rumelhart, G. E. Hinton, and R. J. Williams, "Learning representations by back-propagating errors," *Nature*, vol. 323. pp. 533–536, 1986.

[32] Y. Bengio, "Learning Deep Architectures for AI," *Found. Trends® Mach. Learn.*, vol. 2, no. 1, pp. 1–127, Jan. 2009.

[33] X. Glorot and Y. Bengio, "Understanding the difficulty of training deep feedforward neural networks," *Int. Conf. ...*, vol. 9, pp. 249–256, 2010.

[34] H. Larochelle, Y. Bengio, J. Louradour, and P. Lamblin, "Exploring Strategies for Training Deep Neural Networks," *J. Mach. Learn. Res.*, vol. 10, pp. 1–40, Dec. 2009.

[35] V. Marx, "Biology: The big challenges of big data," *Nature*, vol. 498, pp. 255–260, 2013.

[36] A. H. B. James Manyika, Michael Chui, Brad Brown, Jacques Bughin, Richard Dobbs, Charles Roxburgh, "Big data: The next frontier for innovation, competition, and productivity," *McKinsey Global Institute*, 2011. .

[37] E. McDonald and C. Brown, "khmer: Working with Big Data in Bioinformatics," *arXiv Prepr. arXiv1303.2223*, pp. 1–18, 2013.

[38] K. H. Campbell, J. McWhir, W. A. Ritchie, and I. Wilmut, "Sheep cloned by nuclear transfer from a cultured cell line.," *Nature*, vol. 380, no. 6569, pp. 64–6,

Mar. 1996.

[39] D. E. Cohen and D. Melton, "Turning straw into gold: directing cell fate for regenerative medicine.," *Nat. Rev. Genet.*, vol. 12, no. 4, pp. 243–52, Apr. 2011.

[40] S. J. Morrison, N. M. Shah, and D. J. Anderson, "Regulatory mechanisms in stem cell biology.," *Cell*, vol. 88, no. 3, pp. 287–98, Feb. 1997.

[41] A. G. Smith, "Embryo-derived stem cells: of mice and men.," *Annu. Rev. Cell Dev. Biol.*, vol. 17, pp. 435–62, Jan. 2001.

[42] S. H. Orkin and L. I. Zon, "Hematopoiesis: an evolving paradigm for stem cell biology.," *Cell*, vol. 132, no. 4, pp. 631–44, Mar. 2008.

[43] E. D. Thomas, R. A. Clift, A. Fefer, F. R. Appelbaum, P. Beatty, W. I. Bensinger, C. D. Buckner, M. A. Cheever, H. J. Deeg, and K. Doney, "Marrow transplantation for the treatment of chronic myelogenous leukemia.," *Ann. Intern. Med.*, vol. 104, no. 2, pp. 155–63, Feb. 1986.

[44] F. Candotti, K. L. Shaw, L. Muul, D. Carbonaro, R. Sokolic, C. Choi, S. H. Schurman, E. Garabedian, C. Kesserwan, G. J. Jagadeesh, P.-Y. Fu, E. Gschweng, A. Cooper, J. F. Tisdale, K. I. Weinberg, G. M. Crooks, N. Kapoor, A. Shah, H. Abdel-Azim, X.-J. Yu, M. Smogorzewska, A. S. Wayne, H. M. Rosenblatt, C. M. Davis, C. Hanson, R. G. Rishi, X. Wang, D. Gjertson, O. O. Yang, A. Balamurugan, G. Bauer, J. A. Ireland, B. C. Engel, G. M. Podsakoff, M. S. Hershfield, R. M. Blaese, R. Parkman, and D. B. Kohn, "Gene therapy for adenosine deaminase-deficient severe combined immune deficiency: clinical comparison of retroviral vectors and treatment plans.," *Blood*, vol. 120, no. 18, pp. 3635–46, Nov. 2012.

[45] J. Seita and I. L. Weissman, "Hematopoietic stem cell: self-renewal versus differentiation.," *Wiley Interdiscip. Rev. Syst. Biol. Med.*, vol. 2, no. 6, pp. 640–53, 2011.

[46] M. A. Rieger, P. S. Hoppe, B. M. Smejkal, A. C. Eitelhuber, and T. Schroeder, "Hematopoietic cytokines can instruct lineage choice.," *Science*, vol. 325, no. 5937, pp. 217–8, Jul. 2009.

[47] T. Schroeder, "Long-term single-cell imaging of mammalian stem cells.," *Nat. Methods*, vol. 8, no. 4 Suppl, pp. S30–5, Apr. 2011.

[48] T. Schroeder, "Imaging stem-cell-driven regeneration in mammals.," *Nature*, vol. 453, no. 7193, pp. 345–51, May 2008.

[49] S. G. Megason and S. E. Fraser, "Imaging in systems biology.," *Cell*, vol. 130, no. 5, pp. 784–95, Sep. 2007.

[50] M. Baker, "Cellular imaging: Taking a long, hard look.," *Nature*, vol. 466, no. 7310, pp. 1137–40, Aug. 2010.

[51] U. Rothbauer, K. Zolghadr, S. Tillib, D. Nowak, L. Schermelleh, A. Gahl, N. Backmann, K. Conrath, S. Muyldermans, M. C. Cardoso, and H. Leonhardt, "Targeting and tracing antigens in live cells with fluorescent nanobodies.," *Nat. Methods*, vol. 3, no. 11, pp. 887–9, Nov. 2006.

[52] N. C. Shaner, P. A. Steinbach, and R. Y. Tsien, "A guide to choosing fluorescent proteins.," *Nat. Methods*, vol. 2, pp. 905–909, 2005.

[53] P. S. Hoppe, M. Schwarzfischer, D. Loeffler, K. D. Kokkaliaris, O. Hilsenbeck, N. Moritz, M. Endele, A. Filipczyk1, A. M. Rieger, C. Marr, M. Strasser, B. Schauberger, I. Burtscher, O. Ermakova, A. Bürger, H. Lickert, C. Nerlov, F. J. Theis, and T. Schroeder, "Random PU.1 / Gata1 protein ratios do not induce early myeloid lineage choice," 2014.

[54] K. Manuel, "Using deep neural networks to predict the lineage choice of hematopoietic stem cells from time-lapse microscopy images," 2014.

[55] F. Buggenthin, "Computational prediction of hematopoietic cell fates using single cell time lapse imaging," L, 2011.

[56] W. S. McCulloch and W. Pitts, "A logical calculus of the ideas immanent in nervous activity," *Bull. Math. Biophys.*, vol. 5, no. 4, pp. 115–133, Dec. 1943.

[57] F. Rosenblatt, "The perceptron: a probabilistic model for information storage and organization in the brain.," *Psychol. Rev.*, 1958.

[58] U. of M. LISA lab, "Deep Learning Tutorial," 2014.

[59] Wikipedia, "Sigmoid function," 2014. [Online]. Available: https://en.wikipedia.org/wiki/Sigmoid_function.

[60] *Neural Smithing: Supervised Learning in Feedforward Artificial Neural Networks.* Mit Pr, 1999.

[61] *Introduction to Neural and Cognitive Modeling [Hardcover].* Psychology Press; 2 edition, 2000.

[62] Wikibooks, "Artificial Neural Networks," 2014. [Online]. Available: https://en.wikibooks.org/wiki/Artificial_Neural_Networks.

[63] Y. Bengio, P. Simard, and P. Frasconi, "Learning long-term dependencies with gradient descent is difficult.," *IEEE Trans. Neural Netw.*, vol. 5, pp. 157–166, 1994.

[64] R. Pascanu, T. Mikolov, and Y. Bengio, "On the difficulty of training recurrent neural networks," *arXiv Prepr. arXiv1211.5063*, 2012.

[65] A. Graves, *Supervised Sequence Labelling with Recurrent Neural Networks*, vol. 385. Berlin, Heidelberg: Springer Berlin Heidelberg, 2012.

[66] S. Hochreiter and J. Schmidhuber, "Long short-term memory," *Neural Comput.*, vol. 9, no. 8, pp. 1–32, 1997.

[67] A. Graves, N. Jaitly, and A. Mohamed, "Hybrid speech recognition with Deep Bidirectional LSTM," *... Speech Recognit. ...*, 2013.

[68] A. Graves, M. Liwicki, S. Fernández, R. Bertolami, H. Bunke, and J. Schmidhuber, "A novel connectionist system for unconstrained handwriting recognition.," *IEEE Trans. Pattern Anal. Mach. Intell.*, vol. 31, pp. 855–868, 2009.

[69] M. Schuster and K. K. Paliwal, "Bidirectional recurrent neural networks," *Signal Process. IEEE Trans.*, vol. 45, pp. 2673–2681, 1997.

[70] Y. LeCun, L. Bottou, G. Orr, and K. Muller, *Neural Networks: Tricks of the Trade.* 1998.

[71] I. Sutskever and J. Martens, "On the importance of initialization and momentum in deep learning," ... *Mach. Learn.* ..., no. 2010, 2013.

[72] H. Jaeger and H. Haas, "Harnessing nonlinearity: predicting chaotic systems and saving energy in wireless communication.," *Science*, vol. 304, pp. 78–80, 2004.

[73] G. Hinton, "Neural Networks for Machine Learning," 2012.

[74] J. Martens, "Deep learning via Hessian-free optimization," ... *Conf. Mach. Learn. (ICML-10* ..., 2010.

[75] M. Riedmiller and H. Braun, "RPROP - A Fast Adaptive Learning Algorithm," in *Proceedings of the International Symposium on Computer and Information Science VII*, 1992.

[76] A. y Ng, "Feature selection, L1 vs. L2 regularization, and rotational invariance," in *ICML 2004*, 2004, p. 78.

[77] L. Holmstrom and P. Koistinen, "Using additive noise in back-propagation training.," *IEEE Trans. Neural Netw.*, vol. 3, pp. 24–38, 1992.

[78] G. An, "The Effects of Adding Noise During Backpropagation Training on a Generalization Performance," *Neural Computation*, vol. 8. pp. 643–674, 1996.

[79] G. Hinton and N. Srivastava, "Improving neural networks by preventing co-adaptation of feature detectors," *arXiv Prepr. arXiv* ..., pp. 1–18, Jul. 2012.

[80] C. M. M Schwarzfischer, "Efficient fluorescence image normalization for time lapse movies," 2011.

[81] F. Buggenthin, C. Marr, M. Schwarzfischer, P. S. Hoppe, O. Hilsenbeck, T. Schroeder, and F. J. Theis, "An automatic method for robust and fast cell detection in bright field images from high-throughput microscopy.," *BMC Bioinformatics*, vol. 14, no. 1, p. 297, Jan. 2013.

[82] J. Matas, O. Chum, M. Urban, and T. Pajdla, "Robust wide-baseline stereo from maximally stable extremal regions," in *Image and Vision Computing*, 2004, vol. 22, pp. 761–767.

[83] N. Dalal and B. Triggs, "Histograms of Oriented Gradients for Human Detection," in *2005 IEEE Computer Society Conference on Computer Vision and Pattern Recognition (CVPR'05)*, 2005, vol. 1, pp. 886–893.

[84] von F. Zernike, "Beugungstheorie des schneidenver-fahrens und seiner verbesserten form, der phasenkontrastmethode," *Physica*, vol. 1, no. 7–12, pp. 689–704, May 1934.

[85] H. Tamura, S. Mori, and T. Yamawaki, "Textural Features Corresponding to Visual Perception," *IEEE Trans. Syst. Man. Cybern.*, vol. 8, no. 6, pp. 460–473, 1978.

[86] J. Bayer, C. Osendorfer, D. Korhammer, N. Chen, S. Urban, and P. van der Smagt, "On Fast Dropout and its Applicability to Recurrent Networks," *arXiv1311.0701 [cs, stat]*, 2013.

[87]   S. Wang and C. Manning, "Fast dropout training," *Proc. 30th ...*, vol. 28, 2013.

[88]   V. Pham, C. Kermorvant, and J. Louradour, "Dropout Improves Recurrent Neural Networks for Handwriting Recognition," *arXiv Prepr. arXiv1312.4569*, 2013.

[89]   A. Graves and J. Schmidhuber, "Framewise phoneme classification with bidirectional LSTM and other neural network architectures.," *Neural Netw.*, vol. 18, no. 5–6, pp. 602–10, 2005.

[90]   J. Bergstra, O. Breuleux, F. Bastien, P. Lamblin, R. Pascanu, G. Desjardins, J. Turian, D. Warde-Farley, and Y. Bengio, "Theano: a CPU and GPU Math Expression Compiler," in *Proceedings of the Python for Scientific Computing Conference ({SciPy})*, 2010, pp. 1–7.

[91]   R. Agrawal and R. Srikant, "Fast Algorithms for Mining Association Rules in Large Databases," pp. 487–499, Sep. 1994.

[92]   C. C. Lin, A. Y. Ng, and C. D. Manning, "Parsing Natural Scenes and Natural Language."

Printed in the United States
By Bookmasters